Signals and Communication Technology

More information about this series at http://www.springer.com/series/4748

Obaid Ur-Rehman • Natasa Zivic

Noise Tolerant Data Authentication for Wireless Communication

 Springer

Obaid Ur-Rehman
Data Communications Systems
University of Siegen
Siegen, Germany

Natasa Zivic
Data Communications Systems
University of Siegen
Siegen, Germany

ISSN 1860-4862   ISSN 1860-4870   (electronic)
Signals and Communication Technology
ISBN 978-3-030-07686-3   ISBN 978-3-319-78942-2   (eBook)
https://doi.org/10.1007/978-3-319-78942-2

Printed on acid-free paper

This Springer imprint is published by the registered company Springer International Publishing AG part of Springer Nature.
The registered company address is: Gewerbestrasse 11, 6330 Cham, Switzerland

Preface

With the rapid development and ease in digital content creation and distribution over the Internet, digital data sharing has become a ubiquitous part of our daily lives. This includes data in the form of digital multimedia such as text, audio, video, and images. At the same time, there has been an enormous growth in the methods to compromise the authenticity and authentication methods of data transmitted over modern communication networks as well as the data stored.

Standard authentication mechanisms are very fragile and not robust to any modifications. Even a single bit modification will render the data useless after failing the authentication test at the receiver. Most of the times, a retransmission channel is available and the data can be retransmitted by the source. However, sometimes such a retransmission channel does not exist, e.g., in case of satellite transmission. In such cases, there is a need for data authentication mechanisms to be adopted in such a manner that they are not only able to tolerate the unintentional modifications but also identify, locate, and correct certain modifications if possible.

This is made possible by the use of some recent algorithms for data authentication which cooperate with the techniques for forward error correction, so that minor modifications or errors are not only tolerated but also corrected to the extent possible. This depends on many factors such as the error correction capability of the forward error correction codes used by the noise tolerant authentication method and also the noise tolerance capability of the noise tolerant authentication schemes. It is of course a danger that by allowing more robustness, the security might be compromised. Some of the noise tolerant authentication methods discussed in this book have the ability to configure the robustness using certain parameters of the scheme.

In order to maintain a balance between discussing too much details vs introducing the data authentication mechanisms, this book is organized in seven chapters. The first four chapters introduce the background material needed to understand the next three chapter. The last three chapters are based on the results of the research work conducted by the authors while working at the University of Siegen, Germany.

The salient topics of each chapter in the book are listed below.

Chapter 1 introduces the basics of data authentication in the presence of modifications. Data produced at the source can be modified in many ways from source to sink. This includes intentional and unintentional modifications, where the former ones are more serious but even the latter ones should be given a spatial attention.

Chapter 2 covers the characteristics of a wireless communication channel and how the data transfer is affected by the channel. With the ever-increasing uses of modern gadgets equipped with Internet connectivity, more and more data is being transmitted over the wireless channel in addition to the standard wired channel. Different means to counter the data corruption introduced by communication over a wireless channel are introduced in the chapter.

Chapter 3 discusses the need for robust authentication mechanisms. These authentication mechanisms, as opposed to the standard data authentication mechanisms, are tolerant to certain modifications. Some of these mechanisms are designed to tolerate a few bit modifications, whereas others can additionally also correct the modified data parts by using forward error correction codes as a part of the authentication. Noise tolerant authentication using specially designed message authentication codes as well as digital watermarking is discussed in the chapter.

Chapter 4 introduces the digital watermarking techniques for authentication of multimedia data. This chapter serves as an introduction to the digital watermarking techniques discussed in the following chapters. Digital watermarking, its classification, and characteristics are discussed.

Chapter 5 discusses two digital watermarking techniques for data authentication in the presence of noise. The watermarking techniques discussed in this chapter have the ability of noise tolerant data authentication as well as error correction. The discussed techniques use dual watermarking, where the ability of detection of modified locations is obtained. The additional use of forward error correction codes gives the techniques the ability to correct a few modifications. This is useful if the modifications are introduced unintentionally, e.g., by channel noise.

Chapter 6 covers a method for image authentication using a modified method for image authentication based on standard message authentication codes and channel codes. The discussed method splits the image into important parts, called the region of interest. By authenticating the region of interest using a soft input decryption method and using the authenticated parts as a feedback for decoding the result of channel decoding is also improved in addition to noise tolerant authentication.

Chapter 7, which is the last chapter, discusses two noise tolerant data authentication algorithms which are based on noise tolerant message authentication codes. These codes are themselves based on the standard message authentication codes. The discussed methods have the ability to detect modifications and localize the modifications to block level. A block size can be defined which controls the granularity of modification localization. These algorithms also have the inherent capability of error correction. If the modifications are more than the allowed limits and are beyond the error correction capability of the forward error correction codes, then the modifications are categorized as forgeries.

Siegen, Germany   Obaid Ur-Rehman
                                                                                           Natasa Zivic

Acknowledgments

This book is based on the research work that we, the authors, performed at the chair for data communications systems at the University of Siegen, Germany. We would like to acknowledge the contributions of our former and present colleagues at the chair, as without their support this book would not have been so easy. We would in particular like to sincerely thank the following persons for their support.

Our first and foremost thanks go to Christoph Ruland, who is not only the chairperson of the chair for data communications systems but also has been the doctoral supervisor of both the authors. He has enabled us with the environment needed for working on the research topics discussed in the book. He also supported us by enabling us to work on the international cooperation with the University of Shanghai, China. The research ideas exchanged during the cooperation have formed a basis for certain advanced topics discussed in the book.

Secondly, we would also like to thank our ex-colleague Amir Tabatabaei, who cooperated and worked together with us on multiple noise tolerant data authentication schemes. We thank him especially for extending his support on the security analysis of the ideas.

Last but not the least, we would also like to thank our partners, Chen Ling and Wenjun Zhang, whom we initially got to know during the research project on noise tolerant video authentication with the University of Shanghai. Through various interactions and discussions during the cooperation, we worked together on many interesting ideas in the field of digital watermarking for authentication of video streams.

Most importantly, we would also like to thank our families for their patience as we worked on the book.

Contents

Chapter 1
Introduction and the Need for Noise Tolerant Data Authentication

Abstract In this chapter, the concept of noisy data and its authentication mechanisms are introduced. Standard message authentication codes are good at authenticating data as long as the data or its authentication tags have not changed. However, the techniques for authenticating noisy data, i.e., data which has changed during storage or transmission, are relatively new and not standardized at the time of this writing. The minor modifications in the data could be introduced due to multiple reasons such as storage and transmission errors, especially when the data is transmitted over a wireless medium.

1.1 Noisy Data

Modern communications systems include data capture at the source and transmission to the sink over different communication media. Sometimes the source and sink are directly connected over a wired medium. Most often, the source and sink are farther apart and the communication is performed indirectly over a combination of wired and wireless links. In some cases, the data is stored by the source on a storage medium and later accessed by the sink. In all of these different kinds of storage and communications channels, the data can be influenced by noise resulting in modifications from the original data. Data modified due to any intentional or unintentional reasons are considered and referred to as noisy data in this writing.

Standard data authentication techniques, when applied to noisy data, result in failed authentication. In certain cases, the modifications induced by the storage or transmission media, also known as channel or storage errors, can be repaired using forward error correction codes (FEC). FEC codes are a class of algorithms that try to correct changes in the data by inserting parity at the source and using the parity to detect error locations and magnitudes at the sink. The FEC code then attempts to eliminate those errors. However, the FEC algorithms have a certain error correction capability. If the number of modifications is too high, i.e., beyond the error correction capability of the used FEC code, it might not be possible to eliminate all the

© Springer International Publishing AG, part of Springer Nature 2018
O. Ur-Rehman, N. Zivic, *Noise Tolerant Data Authentication for Wireless Communication*, Signals and Communication Technology,
https://doi.org/10.1007/978-3-319-78942-2_1

modifications using FEC alone. Many different FEC codes have been proposed in the literature and some of them are discussed in Chap. 2.

It is therefore also important to have the ability to authenticate the noisy data as well.

1.2   Data Authentication

Data authentication is composed of two different properties, i.e., data integrity and data origin authentication. Data integrity is defined in [1] as the property that the data has not been altered or destroyed in an unauthorized manner. Data origin authentication ensures that the data comes from the declared source and has not been fabricated by a third party.

Standard method for data authentication (also known as message authentication) uses the well-known message authentication codes (MAC). At the source, a MAC tag (called simply MAC from this point onward) is calculated on the data using a secret key. The MAC is transmitted together with the data to the sink (or otherwise stored together with the data on the storage medium). The sink recalculates the MAC on the data it has received over a communication channel or accessed from the storage medium. If the received and recalculated MAC are the same, the data or message is declared authentic, otherwise it is declared unauthentic. Since a shared secret key is used in calculating the MAC, the data origin authentication is ensured as long as the key is kept secret. Other standard mechanisms which provide data authentication include authenticated encryption and digital signatures. Authenticated encryption schemes not only provide data authentication but also data confidentiality. Digital signature schemes use public key cryptography where the source and sink have different keys. Message authentication codes, on the other hand, use a single key (called a secret key) shared between the source and sink.

1.3   Message Authentication Code

A message authentication code algorithm computes a short string (MAC) as a complex function of every data bit and every secret key bit [2]. A MAC algorithm provides unforgeability in addition to integrity and data origin authenticity. Unforgeability ensures that someone without the knowledge of the secret key should not be able to predict the MAC for new data.

A MAC algorithm must satisfy the following properties:

- For a given key and data, the function can be efficiently computed.
- With no prior knowledge of the key, it is computationally infeasible to find the MAC for a given data, even if a set of data and corresponding MAC are known.

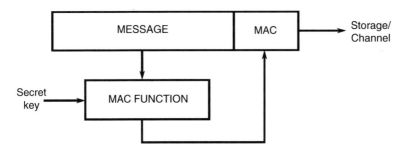

Fig. 1.1 MAC generation

MAC can be generated in a variety of ways. Some of these are standardized by ISO, such as:

- In [3], the methods for generating MAC using block ciphers are standardized.
- In [2], the methods for generating MAC based on dedicated hash functions are standardized.
- In [4], the methods for generating MAC based on a universal hash functions are standardized.

MAC algorithms can also help in the authentication of data origin. Authenticity of data origin [2] means that MAC can provide assurance that a message has been originated by an entity in possession of a specific secret key. Thus, by ensuring the origin, it should not be possible for a third party to fabricate messages on behalf of an original communication partner.

1.3.1   MAC Generation

With the data to be protected, a cryptographic checksum called the MAC tag is calculated on the data with the help of a secret symmetric key and appended to it. The verifier regenerates the MAC tag and compares it with the appended tag to verify the origin as well as integrity of the message. The function for calculating a MAC tag (Fig. 1.1), which is computed over a message M of length m, is a symmetric cryptographic function which results in a MAC of fixed length n, such that $n \leq m$.

Since the MAC generation function reduces the m-bit input to n-bit MAC tag, it is possible that multiple messages have the same MAC tag. Among many applications of MAC codes, the most popular ones are found in banking application, data transmission networks, and industrial applications.

A MAC can be calculated using

- A symmetric block algorithm. In [3] six algorithms for the generation of an m-bit MAC using a symmetric n-bit block cipher are standardized. These include the DES, 3DES, and AES block ciphers. The MAC calculated using this method is

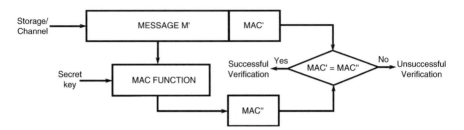

Fig. 1.2 MAC verification

also known as CBC-MAC, where CBC stands for the cipher block chaining mode of operation.
- A dedicated cryptographic hash function, which uses a symmetric key. In [2], three MAC algorithms are standardized that use a secret key and a hash function with an n-bit output to calculate an m-bit MAC.
- A universal hash function initialized using a symmetric key. In [4], the usage of a universal hash function to calculate the MAC tag is standardized. It specifies four MAC algorithms, UMAC, Badger, Poly1305-AES, and GMAC, which use a secret key and a universal hash function with an output size larger than the size of the final MAC.

1.3.2 MAC Verification

The sink gets a possibly modified message and MAC tag pair (M′, MAC′). It recalculates MAC tag on M′ as MAC″ and compares it with MAC′. If MAC″ is equal to MAC′, it is concluded that the data/message is authentic, otherwise the message is declared as unauthentic. In case of a successful match, it is also assumed that the message originated from the sender with whom the receiver shared a secret key (Fig. 1.2).

1.4 Authentication of Noisy Data

The standard techniques for data authentication are good at hard authentication, i.e., declaring data as authentic only when there are no changes in the data and its MAC. Hard authentication mechanisms are not tolerant to any modifications in the data or its MAC. In other words, any changes in the given data are considered as manipulations and the data is declared unauthentic. In most of the standard applications, such a hard authentication is very important for the security of data when no changes are allowed.

However, in the recently emerging multimedia applications, it might not be convenient to have hard authentication. For example, multimedia data such as images, audio, and video may have small modifications and still be perfectly acceptable as authentic. Due to the inherent visual nature of multimedia, in order for data to be considered authentic, it should look or sound authentic, despite minor modifications in the actual bits of the data. However, if the number of modifications is too high, the data might not be recognizable as authentic any more. In this case, it should be possible for the authentication mechanism to detect it. Nevertheless, when there are real forgeries, they should naturally be detectable as well. This calls for the need to look for authentication mechanisms which are a little lenient as compared to the standard hard authentication mechanisms.

The "soft authentication" mechanisms should be able to tolerate minor modifications as opposed to standard "hard authentication" mechanisms but they should still be able to detect real forgeries. They should also be intolerant to large modifications which change the visual representation or meaning of data.

1.5   Data Versus Content Authentication

The soft authentication (also called noise tolerant authentication) mechanisms are sometimes based on the actual data bits. This is also the case for standard hard authentication algorithms, which authenticate the actual data or message bits. Noise tolerant data authentication algorithms are nevertheless mostly based on the data content instead of the actual data.

This makes the authentication more meaningful when the modifications of data do not affect the content. It is to be noted that the modifications do not always have to be unintentional from the perspective of the source. The modifications could also be intentional, such as when quantization, compression, etc. are performed at the source. For example, consider an image which is distorted by minor intentional modifications such as quantization while being processed at the source. Though the quantization might change the data of the image (the actual bit representation), the content of the image might remain unchanged. This might be the case when changing certain bits change only change the intensities of certain pixel colors. However, the content might not be affected by these changes. The content is mostly extracted using feature detection and extraction techniques, which give a footprint of the data.

Noise tolerant data authentication algorithms can also be based on the actual data instead of its content. Most often the methods for nose tolerant data authentication use a threshold to control when to declare the authentication successful. If the allegedly detected modifications are below a threshold, the authentication is declared successful, otherwise the authentication fails.

1.6 Conclusion

In this chapter, a brief overview of the data authentication mechanisms is given. Some standard mechanisms for data authentication are discussed. Thereafter, the importance of having the ability for message authentication in the presence of noise is introduced. The noise tolerant data authentication mechanisms can be based either on data or the content of the data as discussed in the chapter.

References

1. ISO/IEC 7498–2:1989, Information processing systems – Open Systems Interconnection – Basic Reference Model – Part 2: Security Architecture
2. ISO/IEC 9797–2:2011, Information technology – Security techniques – Message Authentication Codes (MACs) – Part 2: Mechanisms using a dedicated hash function
3. ISO/IEC 9797–1:2011, Information technology – Security techniques – Message Authentication Codes (MACs) – Part 1: Mechanisms using a block cipher
4. ISO/IEC 9797–3:2011, Information technology – Security techniques – Message Authentication Codes (MACs) – Part 3: Mechanisms using a universal hash-function

Chapter 2
Wireless Communications

Abstract This chapter focuses on wireless communications, including a discussion of the wireless communication channel and how it impacts the data transmission. The propagation models and fading are discussed further. These phenomena might result in the data to be corrupted in transit from source to sink; some mechanisms must be employed to counter the data corruption. Error concealment methods to counter the problem of data corruption are also discussed in this chapter. They are mainly divided into retransmission techniques and forward error correcting codes. Error correcting codes attempt to correct errors at the sink without a need for retransmission. They are more important from the perspective of this writing and are briefly introduced here. However, their applications are discussed in later chapters.

2.1 Wireless Technology

Wireless communications technology has been around in some form or the other since more than a century. The first public demonstration of radio systems dates back to Marconi's experiments in 1897. This was followed by a rapid progress in the technology, leading to the establishment of radio propagation across the Atlantic Ocean by 1901. Today the most common usage of wireless communications technology is in the form of cellular network, which is rapidly replacing the older wired telephone network.

Wireless communications has made a tremendous impact on our lives by enabling ubiquitous computing. It provides the necessary support for smart homes, smart cities, smart grids and internet of things, to name a few. This enables voice, video, and data traffic with mobility. Interconnecting many different, more often tiny devices has resulted in elimination of the need for thousands of cables. However, due to the nature of wireless communications, far more distortions and disruptions from the environment are experienced as compared to the corresponding wired channel. These disruptions are due to heat, pollution, and physical obstacles in the transmission path, such as houses, buildings, trees, and mountains.

© Springer International Publishing AG, part of Springer Nature 2018   7
O. Ur-Rehman, N. Zivic, *Noise Tolerant Data Authentication for Wireless Communication*, Signals and Communication Technology,
https://doi.org/10.1007/978-3-319-78942-2_2

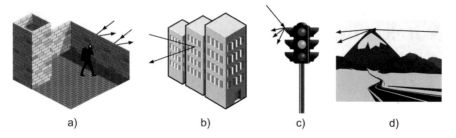

a)   b)   c)   d)

Fig. 2.1 Types of obstacles in wireless communication: (**a**) shadowing (**b**) reflection from big surfaces (**c**) scattering from small surfaces (**d**) diffraction from sharp edges

2.2 Wireless Channel

Certain types of noise are encountered in wireless communication which are usually absent in wired communications channel. A typical example of wireless communications is a mobile network. A mobile communication system consists of a number of base stations (BSs). The base stations are typically stationary, each defining a cell or more precisely the coverage area of the BS. Each base station provides services to mobile stations (MSs) which are present inside the cell.

An MS may or may not have a direct line of sight (LoS) connection to the BS. This might be due to natural hindrances, such as mountains and trees, or human made obstacles, such as houses, buildings, and poles. Thus, there might be one or more non–line of sight (NLoS) paths between an MS and a BS, though they might be in the same cell. Due to these obstacles, the signal from an MS might reach a BS through an indirect LoS radio propagation path. The signal might meet multiple obstacles during the transmission from source to sink. This results in the phenomena such as shadowing, scattering, reflections, refractions, and diffractions (Fig. 2.1).

Shadowing occurs when the direct communication between a source and destination is obstructed by a thick object through which the waves cannot penetrate and do not reach the receiver directly.

Scattering occurs when a ray is incident on small surfaces and as a result the wave is deflected in many different directions.

Reflection occurs when a ray of heat, light, sound, etc. is incident on a surface and a part of it is sent back into the same medium from where it came. The reflected ray has the same angle as that of the angle of incidence.

Refraction occurs when a ray incident on a surface moves from one medium to another. The refracted ray has a different velocity and takes a different angle in the new medium than before.

Diffraction occurs when a ray is incident on a surface with sharp edges and the wave slits or bends around the edge.

Fig. 2.2 Fast and slow
fading

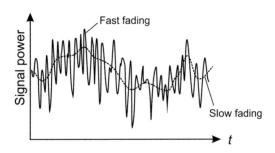

Not only a signal transmitted over a wireless channel is distorted through these obstacles, it is also additionally influenced by transmission noise and disturbances due to other signals. Multiple copies of the same signal might arrive from many different directions at the receiver, where each copy experiences its own level of attenuation, delay, and phase shift. The receiver might receive multiple copies of the signal, which are out of phase or have different amplitudes. In order to reconstruct the original signal, the different copies of the signal need to be combined at the receiver [1–3].

Fading and interference are two major phenomena resulting from wireless transmission.

Fading is the fluctuation in the attenuation of a signal with time, position, frequency, etc. Fading is also defined as a fluctuation in the amplitude or phase of a signal [4]. As the signal traverses from transmitter to receiver, the signal may follow different paths and might get scattered, reflected, or diffracted. As a result the receiver sees multiple copies of the signal, each arriving with a different phase shift, at a different time, or with different amplitude. When the copies of signals are superimposed at the receiver, the result is either a constructive or a destructive interference, i.e., the signal is either amplified or attenuated.

Fading can be further subdivided into fast fading and slow fading.

Fast fading occurs when partial signals arriving at the receiver add up destructively due to time delays and phase shift resulting in short breaks or fades in the amplitude of the resultant signal.

Slow fading occurs due to the changing distance between the transmitter and the receiver. The distance as well as the signal environment due to the physical obstacles leads to slow changes of the average power of the arriving signal. This change in average power over a longer period of time is known as slow fading.

Fast and slow fading of a signal are depicted in Fig. 2.2.

Interference is the superposition of multiple radio waves to produce a resultant wave, which has a higher, lower, or the same amplitude as the original wave. The waves of a signal can either add up constructively or destructively, depending on the relative phase difference between them. The constructive or destructive addition of the waves results in the change in the amplitude of the resulting signal as shown in Fig. 2.3.

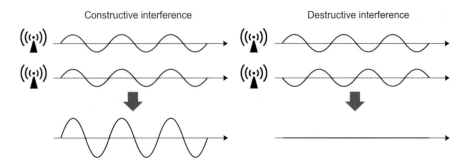

Fig. 2.3 Interference

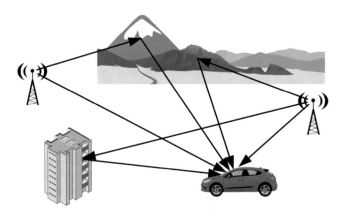

Fig. 2.4 Multipath propagation

2.3 Inter-symbol Interference

In some cases, there is a line of sight (LoS) channel between a transmitter and receiver, such as satellite to ground station communication. In other cases, an LoS channel might not be available, such as mobile telephony. However, one or more non–line of sight (NLoS) paths might be present between a transmitter and receiver. The NLoS paths are a result of the phenomena such as scattering and diffraction as discussed earlier. The obstacles between transmitter and receiver, such as buildings, trees, mountains, make the signal arrive at the receiver through multiple different paths. This phenomena is known as multipath propagation. This is shown in Fig. 2.4, where in addition to one LoS path there are also multiple indirect NLoS paths.

The scattering, reflection, refraction, or diffraction by the obstacles in the signal's path results in multiple copies of the signal arriving at the receiver with varying time delays and different levels of attenuation and phase shifts. The strength of the resultant signal at the receiver depends on the strength of the individual component signals received from multiple paths.

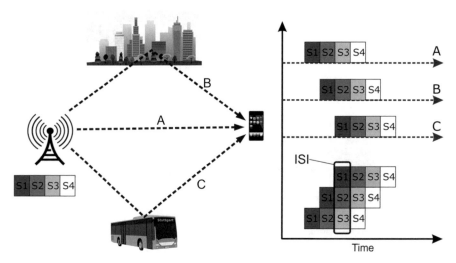

Fig. 2.5 Inter-symbol interference (ISI)

ISI occurs when multiple copies of a signal are received at different times, resulting in a symbol interfering with other symbols, thereby creating an effect of noise. This makes the recovery of information very difficult. Figure 2.5 shows three copies of a signal that are received via paths A, B, and C and are time delayed. The symbols S1, S2, and S3 overlap at the receiver, resulting in ISI. ISI can be improved using error correcting codes, by separating symbols in time with guard periods or by using adaptive equalize. The idea of using error correcting codes is discussed in this book.

2.4   Propagation Model

2.4.1   One-Way Propagation

A signal $r(t)$ (received at time t) is uniquely identified by its amplitude A, its frequency $f_0 = \omega_0/2\pi$, and its phase shift φ_0 [3]:

$$r(t) = Ae^{j(\omega_0 + \varphi_0)} \tag{2.1}$$

In (2.1) it is assumed that the receiver is nonstationary relative to the transmitter and experiences no reflections of the transmitted signal. The phase shift of the received signal varies with the movement. If the distance between the transmitter and receiver changes, the transmitted signal experiences phase change. This change in phase is proportional to the path change at the receiver. For simplicity, it is assumed here that the change in path is small enough so that the average path loss is not affected by the change. The propagation loss is therefore approximately constant. One way propagation is shown in Fig. 2.6.

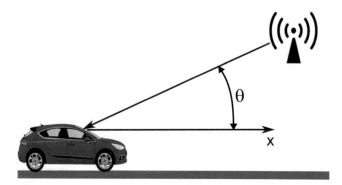

Fig. 2.6 One-way propagation

The signal $r(t)$ is represented as,

$$r(t) = Ae^{j(2\pi f_0 t - \beta x \cos\theta)} \tag{2.2}$$

where β is the wave number such that $\beta = 2\pi/\lambda$.

If the receiver moves in the direction x with the speed v covering a distance $v \cdot t$ in time t, then,

$$r(t) = Ae^{j2\pi t\left(f_0 - \frac{v}{\lambda}\cos\theta\right)} \tag{2.3}$$

where the term $v/\lambda \cdot \cos\theta$ represents the Doppler's frequency f_D. The envelope of the received signal is a constant value given by

$$|r(t)| = A \tag{2.4}$$

The reception level is almost unaffected by small path changes resulting in no fading effect. This is expected, as only one propagation path exists, resulting in no interference of signals.

2.4.2 Two-Way Propagation

If there is a reflection path between the transmitter and receiver, the received signal is made up of two components, i.e., a directly and indirectly received signal.

The propagation loss is of the same magnitude in both copies of the received signals, so the received signal $r(t)$ can be written as [4] (ignoring phase shifts on reflection points):

$$r(t) = \frac{A}{2}e^{(j2\pi f_0 t)}e^{-j2\pi\frac{vt}{\lambda}\left(\frac{\cos\theta_1 + \cos\theta_2}{2}\right)}\left[e^{-j2\pi\frac{vt}{\lambda}\left(\frac{\cos\theta_1 - \cos\theta_2}{2}\right)} + e^{+j2\pi\frac{vt}{\lambda}\left(\frac{\cos\theta_1 - \cos\theta_2}{2}\right)}\right] \tag{2.5}$$

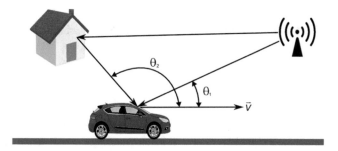

Fig. 2.7 Two-way propagation

The envelope of the received signal is given by (2.6)

$$|r(t)| = A \left| \cos\left[2\pi \frac{vt}{2\lambda}(\cos\theta_1 - \cos\theta_2) \right] \right| \tag{2.6}$$

It can be observed from (2.6) that the envelope of the two-way propagation signal is dependent on the path and the angle of arrival of the signal components and therefore it is not a constant anymore (Fig. 2.7).

2.4.3 N-Way Propagation

In general, a number of copies of the signal (waves) may arrive at the receiver from different directions (θ_i) and with different amplitudes (a_i). Let this number be represented by N. The resultant received signal is derived from the vector superposition of all of these individual waves resulting in

$$r(t) = \sum_{N}^{i=1} a_i e^{(j2\pi f_0 t)} e^{(-j\beta vt \cos\theta_i)} \tag{2.7}$$

2.5 Fading Model

Due to multipath propagation, the waves of signals arrive at the receiver through multiple paths and are typically out of phase. A reduction of signal strength is experienced when there is destructive interference due to wave cancellation. This leads to fading, depending on the amount by which the signals are attenuated, delayed, or phase shifted, resulting in fluctuations in the signal amplitude with time. Due to the presence of multiple paths between a transmitter and receiver, fading may vary with time and the location of the receiver, especially when the receiver is not stationary. Fading is often modeled as a random process as the faded signal's amplitude varies with time and appears to be a random variable at the receiver. There are two main models for fading known as the Rayleigh fading and the Rician fading.

2.5.1 Rayleigh Fading

Let us say a receiver receives a large number of reflected and scattered waves and there is no direct LoS signal reception due to obstacles or the distance between the sender and receiver. Due to the destructive interference resulting in wave cancellations, the instantaneous received signal power appears to be a random variable at the moving receiver. The fading effect experienced by the receiver is called Rayleigh fading, and Rayleigh distribution is used to model the multiple paths of the densely scattered signals reaching the receiver without an LoS component.

The probability density function (PDF) of a Rayleigh distribution is given by

$$f_A(A) = \frac{A}{\sigma^2} e^{-\frac{1}{2}\frac{A^2}{\sigma^2}}, A > 0 \tag{2.8}$$

where A is the absolute value of the amplitude and σ is the shape parameter. The mean and variance of the Rayleigh distribution are given by $E\{A\} = \sigma\sqrt{\frac{\pi}{2}}$ and $\sigma^2\left(2 - \frac{\pi}{2}\right)$, respectively.

2.5.2 Rician Fading

Rician fading is similar to the Rayleigh fading but as opposed to Rayleigh fading, a dominant component of the signal is received by the receiver in Rician fading. This dominant component may be due to the presence of a direct LoS channel. The Rician fading is modeled using a Rician distribution, whose probability density function is given by,

$$f_A(A) = \frac{A}{\sigma^2} e^{-\frac{1}{2}\frac{A^2+s^2}{2\sigma^2}} I_0\left(\frac{As}{\sigma^2}\right) \tag{2.9}$$

where s is the dominant component of the signal and $I_0(x)$ is the modified zero-order Bessel function of the first kind given by

$$I_0(x) \triangleq \frac{1}{\pi} \int_0^\pi e^{x\cos\theta} dx \tag{2.10}$$

If there is no sufficiently strong signal component, s, i.e., if $\frac{s}{\sigma} \to 0$, then the PDF of the Rician distribution equals the PDF of the Rayleigh distribution.

2.6 Doppler Shift

Aside from fading, there are many more effects which influence the wireless transmission. One of the very common effects observed in the case of a moving sender or receiver is the Doppler shift or the Doppler's effect. Doppler shift represents the case of change in the frequency of a wave when the receiver moves relative to the source. With a change in the distance between a source and receiver, the phase of the received signal changes as well.

As the distance between the source and receiver decreases, e.g., when the source moves toward the receiver, each successive wave crest is emitted closer to the receiver as compared to the previous one. This results in the waves bunching together. On the other hand, when the distance between the source and receiver increases, e.g., when the source moves away from the receiver, each successive wave is emitted at a farther distance than the previous. This results in the waves spreading out.

Let a mobile receiver move with a velocity v from point A to point B. During a time Δt, the receiver covers a distance $d = v \cdot \Delta t$. This results in the change of path length and the reception phase at the receiver. As a consequence, the wave from sender to receiver experiences a shift in the frequency. This shift in signal's frequency due to the motion is called the Doppler shift. The Doppler shift is given by

$$f_D = \frac{v}{\lambda} \cos \theta \qquad (2.11)$$

where λ is the wavelength of the signal and θ is the receiving angle.

The maximum Doppler shift is $f_m = \frac{v}{\lambda}$. The two extreme values of the Doppler shift are observed when a mobile station moves directly toward a base station or moves away from it.

Doppler effect is depicted in Fig. 2.8 for a moving source of sound.

2.7 Error Concealment

The errors and distortions introduced by multipath fading can be concealed in a number of ways. Two standard mechanisms to achieve error concealment include forward error correction (FEC) codes and diversity techniques.

2.7.1 Automatic Repeat Request

Automatic repeat request (ARQ) is an error control mechanism that makes use of error detection codes, such as CRC codes, for error detection, and retransmissions based on negative acknowledgments or timeouts. The core idea is that the receiver

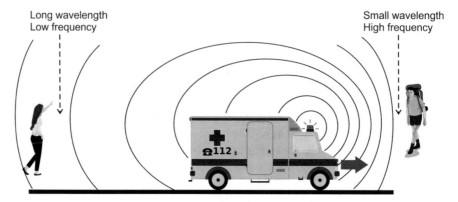

Fig. 2.8 Doppler effect for a moving source of sound

is able to detect errors and then request retransmission from the transmitter if errors are found in the received data.

Retransmission techniques based on ARQ are further divided into multiple categories such as Stop-and-Wait ARQ, Go-Back-N ARQ, and Selective Repeat ARQ.

Stop-and-Wait In Stop-and-Wait ARQ, the transmitter sets a timer before transmitting a packet and then waits for an acknowledgment (ACK) before sending the next packet. If an ACK is not received and a timeout occurs at the transmitter, the packet is retransmitted. The throughput of this approach is very low as most of the available bandwidth is wasted as transmitter keeps waiting for an ACK.

Go-Back-N In Go-Back-N ARQ, the transmitter retransmits N packets if an ACK is not received and timeout occurs. The throughput achieved by this approach is higher than Stop-and-Wait ARQ, but in Go-Back-N, N packets must be kept by the receiver in memory for retransmission when needed, although a few might already have been received by the receiver.

Selective Repeat In Selective Repeat ARQ, only the packets identified by the receiver are retransmitted. As the transmitter only retransmits the requested packets a better throughput is achieved as compared to the other two ARQ techniques discussed above.

2.7.2   Forward Error Correction Codes

Forward error correction (FEC) codes, also known as error correcting codes or channel codes, are techniques for correcting errors at the receiver without the need for requesting a retransmission. The error correction is done based on parity attached by the transmitter to the data. At the receiver, the presence of errors is detected and the number of errors and their locations are found. If the number of errors is below the error correction capability of the code, the errors are corrected. Different error correcting codes are used in practice, such as Reed-Solomon (RS) codes [6], low-density parity-check (LDPC) codes [7], and turbo codes [8].

RS codes are algebraic FEC codes based on Galois field (GF). They are suitable for medium sized message protection. If $GF(2^8)$ is used, then each symbol is 8 bits long. RS code is typically written as RS(n, k), where n is the codeword length, k is the data length, $n-k$ is the number of parity symbols, and k/n is the code rate. Thus, an RS(255, 239) code means that the codeword length is 255 symbols, out of which 239 are data symbols and 16 are parity symbols. The error correction capability of this code is 8 symbols, i.e., $(n-k)/2$. In addition to sporadic error correction, RS codes are good at correcting burst errors. Most widely used hard decision decoding algorithms for Reed-Solomon codes are Berlekamp-Massey [9] algorithm and the Euclidean algorithm [10]. Soft decision decoders for RS codes have also been proposed, which extend the error correction capability beyond that achievable with hard decision decoders. Soft decision decoding is beyond the scope of this text.

Turbo codes were proposed by Berrou, Glavieux, and Thitimajshima in 1993. They are the first practical codes shown to approach the channel capacity. In theory, they are a concatenation of two codes connected by an interleaver. Practically, the constituent codes are almost always convolutional codes connected in parallel. Encoding is performed by interleaving the input data and passing it through the encoders. The input data is passed through an identity interleaver followed by the first encoder and in parallel the input data is passed through another nonidentity interleaver followed by the second encoder. The decoding for turbo codes is usually performed by a soft decision iterative decoder. The task of the decoder is to find the most probable codeword based on the demodulated received sequences. The a priori knowledge or probabilities about the messages and their occurrence are taken into account by the decoder. The decoding is performed by two decoders, each one capable of working on soft input and producing soft output, i.e., the likelihood of the output bits. The most common SISO decoding algorithms are Soft Output Viterbi Algorithm (SOVA) and the Maximum A posteriori Probability (MAP) decoding algorithm.

Low-density parity-check (LDPC) codes are linear block codes, invented by Gallager in his PhD thesis in 1960 [7]. Since their encoding and decoding were too complex for the technology available at that time, they remained largely ignored for the next 30 years. However, Reed-Solomon codes, with their excellent algebraic structure and ease of use and implementation together with their excellent error and erasure correction capability, remained in practice for those years. During these years, Tanner did a remarkable work on LDPC codes in 1981 [11], where he introduced the graphical representation of the LDPC codes. This is now known as the Tanner graph representation. After the introduction of turbo codes and the demonstration of their excellent performance, there was a quest for further capacity approaching codes. This led to the reinventions of LDPC codes in 1993. This is credited to MacKay [12, 13] and Luby [14]. LDPC codes are widely used in many standards such as WiFi/IEEE 802.11 [15, 16], WiMAX/IEEE 802.16e [17], DVB-S2 [18], Ethernet 10GbaseT, etc.

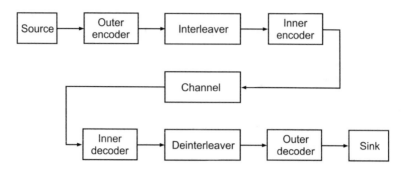

Fig. 2.9 Code concatenation

Sometimes multiple codes are combined together to increase the error correction capability beyond the capability of a single code. This method is called code concatenation where each encoder adds its own parity to the data. Codes are concatenation either in serial or in parallel, although serial code concatenations are used more often. Although the same code can be concatenated, most often different codes are concatenated. When codes are concatenated in serial, the first encoder through which data is encoded is called the encoder for the outer code and the next one as the encoder for the inner code. The decoders are applied in reverse order on the received data, i.e., the decoder for inner code is applied on the received data followed by the decoder for outer code. Code concatenation improves the overall error correction capability and thus the coding gain. Examples of code concatenation are found in many practical applications and standards such as DVB-S2 and compact discs. In Digital Video Broadcast (DVB-S2), LDPC code is used as the inner code and an algebraic code, such as RS code, as the outer code. Compact discs use a concatenation of two Reed-Solomon codes of different lengths with an interleaver placed in between them. The purpose of interleaver/de-interleaver is to spread the sporadic burst of errors to single errors. Code concatenation is shown in Fig. 2.9.

2.7.3   Diversity and Combining Techniques

Diversity methods are used to combat the detrimental effects of channel fading experienced in wireless communications. Multiple copies of a signal are transmitted over statistically independent fading channels. Signals undergo different affects by the different channels and the probability of experiencing simultaneous deep fades by the received copies of the signal is too low. At the receiver, the multiple copies are combined such that the effects of fading are minimized. The receiver diversity mitigates the fluctuations due to fading, which makes the channel appear like an additive white Gaussian noise (AWGN) channel. The probability of experiencing a deep fade in different copies of the signal is very low. Combining multiple copies reduces the effect of fading.

The diversity combining schemes discussed next include selection combining, switched combining, maximal ratio combining, and equal gain combining [5].

2.7.3.1   Selection Combining

In selection combining, the signal with the highest signal to noise ratio (SNR) out of the N received signals is selected. The combiner chooses the signal on the branch with the highest SNR.

2.7.3.2   Switched Combining

Selection combining is very costly to implement as M separate receivers/antennas are required. In switched combining only one receiver is needed, where switching between the antennas is carried out prior to the receiver. The signal whose SNR is above a predetermined threshold is connected through to the receiver. If the SNR of the signal falls below the threshold, the receiver is switched to the second signal regardless of the instantaneous SNR of this signal.

2.7.3.3   Maximal Ratio Combining

In selection combining and switched combining, the signal received on one of the branches is chosen. In maximal ratio combining, the output of the combiner is a weighted sum of all the signals received. Each of the received signal is weighed with a gain factor proportional to its own SNR and phase corrected. All of the weighed branch signals are summed. Although the resultant signal is better than selection combining and threshold combining methods, it is still the most expensive way of combining faded signals.

2.8   Multiple Input Multiple Output

Multiple input multiple output (MIMO) is a wireless transmission technique which uses multiple transmitting and receiving antennas. MIMO systems are more efficient as compared to the corresponding single input single output (SISO) systems due to the higher data rate and lower bit error rate (BER) which implies a higher spectral efficiency and quality of service (QoS). MIMO is presently used in a variety of wireless standards such as Wi-Fi, HSPA+ [19], WiMAX, and LTE [20]. Unlike conventional wireless systems which suffer from the multipath propagation, MIMO systems exploit it for multiplying the capacity of the radio link. The transmitter sends multiple streams by multiple transmit antennas through a matrix channel consisting of all paths between n transmit antennas and m receive antennas.

2.9   Conclusion

In this chapter, the characteristics of wireless communication channel are briefly discussed. Multiple phenomena such as inter-symbol interference, multipath propagation, and fading are experienced in wireless communication due to the nature of wireless transmission channel. These might result in the corruption of data from source to sink. Different solutions are proposed in literature to address these problems. These include mainly retransmission techniques and forward error correction codes. These techniques are briefly introduced in this chapter.

References

1. M. Schwarz, *Mobile Wireless Communications*. The Edinburgh Building, Cambridge CB2 8RU, UK (Cambridge University Press, 2005)
2. J.G. Proakis, *Digital Communications*, 4th edn. New York, US (McGraw-Hill, 2001)
3. C.Y. Lee, *Mobile Communications Engineering-Theory and Applications*, 2nd edn. New York, US (McGraw-Hill, 2008)
4. K. David, T. Benker, *Digitale Mobilfunksysteme*. Berlin, Germany (Teubner, 1996)
5. B. Sklar, Rayleigh fading channels in mobile digital communication systems part I. IEEE Commun. Mag. **35**(7), Jul (1997)
6. I.S. Reed, G. Solomon, Polynomial codes over certain finite fields. J. SIAM Appl. Math. **8**, 300–304 (1960)
7. R. Gallager, Low-Density Parity-Check Codes. PhD thesis, Massachusetts institutes of Technology, 1960
8. C. Berrou, A. Glavieux, P. Thitimajshima, Near Shannon Limit Error Correcting Coding and Decoding: Turbo-Codes. 1, IEEE International Conference on Communications (ICC '93), May 23–26, 1993, Geneva, Switzerland
9. E.R. Berlekamp, Algebraic Coding Theory (McGraw-Hill, New York, 1968), (Revised edition, Laguna Hills: Aegean Park Press, 1984)
10. Y. Sugiyama, Y. Kasahara, S. Hirasawa, T. Namekawa, A method for solving key equation for Goppa codes. Inf. Control. **27**, 87–99 (1975)
11. R.M. Tanner, A recursive approach to low complexity codes. IEEE Trans. Inform. Theory **27**(5), 533–547 (September 1981)
12. D. MacKay, R. Neal, Good codes based on very sparse matrices. Cryptography and Coding, 5th IMA Conference, LNCS, pp. 100–111, Berlin, October 1995
13. D. MacKay, Good error-correcting codes based on very sparse matrices. IEEE Trans. Inform. Theory **45**(2), 399–431 (1999)
14. N. Alan, M. Luby, A linear time erasure-resilient code with nearly optimal recovery. IEEE Trans. Inform. Theory **47**(6), 1732–1736 (1996)
15. 802.11n – IEEE Standard for Information technology – Local and metropolitan area networks - Specific requirements – Part 11: Wireless LAN Medium Access Control (MAC) and Physical Layer (PHY) Specifications Amendment 5: Enhancements for Higher Throughput, 2009
16. 802.11ac – IEEE Standard for Information technology – Telecommunications and information exchange between systems – Local and metropolitan area networks – Specific requirements – Part 11: Wireless LAN Medium Access Control (MAC) and Physical Layer (PHY) Specifications—Amendment 4: Enhancements for Very High Throughput for Operation in Bands below 6 GHz,2013

17. 802.16 – IEEE Standard for Local and metropolitan area networks Part 16: Air Interface for Broadband Wireless Access Systems, 2009
18. "Digital Video Broadcasting (DVB): Second generation framing structure, channel coding and modulation systems for Broadcasting, Interactive Services, News Gathering and other broadband satellite applications (DVB-S2)", ETSI EN 302 307, April 2009
19. 3GPP release 7: High Speed Packet data Access (HSPA), 2011
20. 3GPP TS 36.201: Evolved Universal Terrestrial Radio Access (E-UTRA); physical layer; general description, Release 13, 2016

Chapter 3
Noise Tolerant Data Authentication Mechanisms

Abstract In this chapter, some important noise tolerant data authentication mechanisms are discussed. These include the generic noise tolerant data authentication constructs as well as the one built specifically for content-based authentication. Some of the noise tolerant data authentication algorithms internally use forward error correcting codes to provide the additional ability of error location and correction in addition to noise tolerant authentication. These techniques are also briefly discussed in this chapter. Noise tolerant data authentication using digital watermarking is also quite common, and a brief introduction to such watermarking techniques is given in this chapter as well.

3.1 Approximate Message Authentication Code

Approximate message authentication code (AMAC) is a special kind of message authentication code (MAC) which, unlike the standard MAC, is tolerant to minor modifications in the message as long as the number of modifications is below a certain threshold. The threshold can be predefined and is adjustable using the parameters of the code. AMAC was first proposed in [1]. The error tolerance in AMAC is achieved using majority logic. The AMAC tag (A) computed on two messages M and M', such that M $\neq$ M', might be the same. Here M and M' should ideally have only minor differences. This gives AMAC the ability to tolerate modifications in the protected data.

AMAC tag generation is shown in Fig. 3.1. Let L be the length of the AMAC tag. Let R and S be positive integers used to split the message into rows and columns and to influence the majority logic to calculate the tag. Thus, the parameters L, R, and S set the threshold for the generation of AMAC tag. At first the message M is padded with zeros so that it is $L{\cdot}R{\cdot}S$ bits long. The message is arranged in rows and columns, such that the number of columns is L and the number of rows is $R{\cdot}S$. A pseudorandom bit permutation is applied on this arrangement of rows and columns to get a permuted matrix. The bit permutation needs to be repeatable at the receiver, so

© Springer International Publishing AG, part of Springer Nature 2018
O. Ur-Rehman, N. Zivic, *Noise Tolerant Data Authentication for Wireless Communication*, Signals and Communication Technology,
https://doi.org/10.1007/978-3-319-78942-2_3

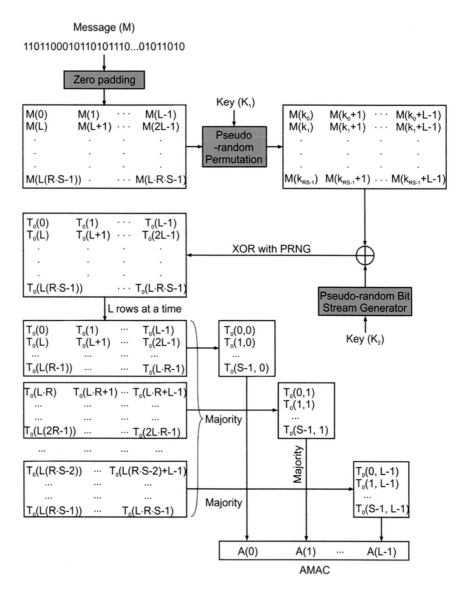

Fig. 3.1 AMAC tag generation [1]

a shared secret key can be used as a seed to obtain the permutation sequence. The resultant data is then XORed with a pseudorandom bit stream. The pseudorandom number generator is initialized with a shared key as a seed. The key is pre-shared between the sender and the receiver using standard key sharing mechanisms. The output of the XOR is split into S groups of R rows, where the majority logic is used to select the output bits of the AMAC tag. The majority helps in obtaining the tolerance to modifications. When the number of modifications does not change the majority of a bit in rows or columns dictated by L, R, and S, then the same AMAC is obtained for the different data.

A security analysis of AMAC is given in [2], where it was shown in [2] that AMAC is vulnerable to a security attack. A solution to counter this vulnerability was also proposed in [2]. The solution involves using two pseudorandom bits XORed with the message bit instead of one pseudorandom bit. The modified AMAC is referred to as the strengthened AMAC in this text.

Application of AMAC in image authentication was demonstrated through [3] in what is known as the approximate image message authentication code (AIMAC).

3.2 Soft Input Decryption for Data Authentication

Soft input decryption (SID) algorithm for data authentication was proposed in [4]. A standard decryption algorithm decrypts the given data with the help of a key, without any other side information. Soft input decryption internally uses an iterative decoding and decryption procedure, where a standard decryption algorithm is used together with additional side information from a channel decoder such as convolutional codes or the channel measurements directly. The channel measurements or bit reliabilities provided by the channel decoder such as the decoder for convolutional codes or turbo codes [5] are used by the SID algorithm to authenticate data. These bit reliabilities are also known as L-values or log-likelihood ratios (LLRs) [6].

A soft input soft output (SISO) channel decoder, such as a decoder for convolutional codes or turbo codes, produces an LLR value for each decoded bit. The L-values or the LLRs are a measure of reliability of the bits decoded by the channel decoder. A higher absolute value of the LLR of a bit means a higher probability that the bit is correctly decoded. Vice versa, a lower absolute LLR value of a bit means a lower probability that the bit is correctly decoded. Though typically, the LLRs can be obtained from a decoder, such as the decoder for convolutional or turbo codes. However, in the absence of channel codes, the LLR values can be directly obtained by observing the channel measurements. These are called the reliability values or L-values of the channel. Let a codeword c of length n ($c = (c_1, c_2, c_3, c_4, \ldots, c_n)$) be transmitted over a channel. The SISO channel decoder produces the LLR of each decoded bit as

$$\text{LLR}\left(c_i\right) = \log\left(\frac{P\left(u_i = 1\right)}{P\left(u_i = 0\right)}\right), i = 1, 2, 3, \ldots\ldots, n \qquad (3.1)$$

SID is an iterative algorithm for correction and authentication of a message and its authentication tag, e.g., a MAC tag. The iterative decryption and decoding by flipping a combination of least reliable bits helps in correction of minor errors in the message and/or its authentication tag. In practice, a channel decoder is already used by almost all of the standard communication protocols. Thus most of the errors are already corrected by the channel decoder. If even a single error is left over by the channel decoder, then the standard authentication methods would fail to authenticate the data. SID for authentication is very good at correcting the minor left over errors and authenticating the data at the same time. However, if the number of errors exceeds a threshold, this method will declare the data un-authentic.

SID uses the LLRs to improve the decoding and decryption performance. The bits of the data and its authentication tag are sorted based on their reliability values. A combination of least reliable bits is chosen in each iteration and these bits are flipped, analogous to the Chase decoder [7]. After flipping the chosen combination of bits, a standard decryption or authentication algorithm is run on the modified data and tag pair. If the authentication fails, the next combinations of bits are selected and flipped. The procedure is repeated till the authentication succeeds. The iterative decoding is stopped if a predefined number of iterations (the threshold) have been performed. In the simplest form, the maximum number of iterations of SID, N_{max}, is dictated by the number of least reliable bits to be considered for correction. It is represented as two to the power of the number of least reliable bits to be considered for correction (b), i.e.,

$$N_{max} = 2^b \qquad (3.2)$$

A typical value of b is taken as 8, 16, or 24. The noise tolerant authentication capability of SID increases with the higher value of b. However, larger value of b also means more iterations of SID and therefore the need for higher processing power. If authentication does not succeed, even after N_{max} iterations, the data is declared as unauthentic. The SID algorithm works well for smaller data sizes. The algorithm of SID at the receiver is shown in Fig. 3.2.

SID improves the results of decryption and of channel decoding at the same time. This is due to the fact that it is able to correct minor errors induced during transmission over a noisy communication channel.

3.3   Noise Tolerant Message Authentication Code

The noise tolerant message authentication code (NTMAC) method [8] is based on the standard MAC algorithms. The message to be authenticated is split into blocks and a standard MAC is calculated over each block. For each block, NTMAC retains a part of the complete MAC, called the sub-MAC. All the sub-MAC tags are appended together to form the NTMAC tag of the entire message.

During authentication, the process of NTMAC tag calculation is repeated for the received message. For an unaltered message, all the received sub-MACs should match the recalculated sub-MACs. A threshold can be chosen and used to determine the number of recalculated sub-MACs that must match the received sub-MACs for noise tolerant authentication to succeed.

NTMAC tag generation is shown in Fig. 3.3.

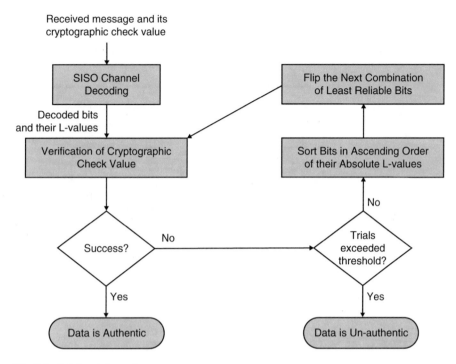

Fig. 3.2 Soft input decryption

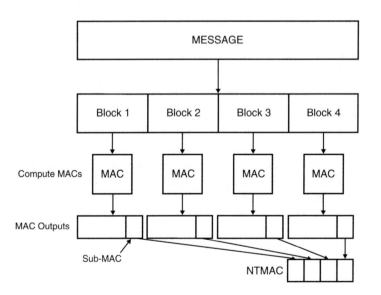

Fig. 3.3 NTMAC tag generation

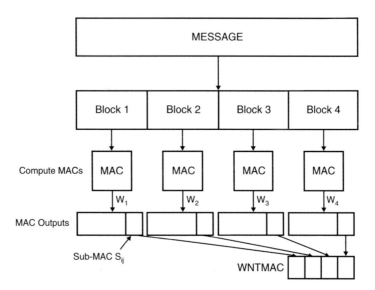

Fig. 3.4 WNTMAC tag generation

3.4 Weighted Noise Tolerant Message Authentication Code

Weighted NTMAC (WNTMAC) [9] is an extension of the NTMAC algorithm. The concept of relative weights is introduced in WNTMAC. A weight is assigned to each message block (and therefore the corresponding sub-MAC). A higher weight is assigned to the more important part of the data to be protected with NTMAC and a lower weight is assigned to the relatively lesser important part of data. A higher weight essentially means higher protection using a larger sub-MAC, whereas a lower weight means relatively lesser protection, i.e., a smaller sub-MAC and essentially higher chances of collision and forgery.

The application of WNTMAC was demonstrated in images in [10]. Image to be protected is split into rows and columns of nonoverlapping blocks. Fourier transform is calculated on the complete image. The higher frequency elements of the Fourier transform are more important than the low frequency elements and also very essential for image recover. Therefore, the DC element and the first few AC elements are assigned a relatively higher weight than the low frequency elements. The data assigned higher weight is assigned a larger sub-MAC. The individual data blocks are decoded at the receiver in the same manner as in SID. The data blocks of higher weight are also assigned higher iterations than the data blocks of lower weights. Thus, the chances of recovery of data blocks of higher weight are higher than the chances of recovery of data blocks of lower weights. It is demonstrated in [10] that the bit error rate of the individual message blocks of higher weight is better than the bit error rate of other message blocks.

WNTMAC tag generation is shown in Fig. 3.4.

3.5   Authentication Based on Feature Extraction

Authentication mechanisms based on feature extraction techniques are also very popular. These methods are even more popular in multimedia authentication. Features are a footprint of the data, which can be authenticated rather than the data itself. Since a change in data might keep the features intact, noise tolerant authentication can be achieved. There are various feature extraction techniques as proposed in literature. The most basic feature extraction methods consider the Fourier transform or cosine transform of given data (image, voice video, etc.). More advanced feature extraction techniques include histogram extraction, edge detection, and scale invariant feature transform (SIFT) [11].

Features can be extracted on the whole data (such as an image) or parts of the data. When feature extraction and the corresponding authentication are performed on the whole data, noise tolerant authentication can be performed. However, error location is more difficult in this case. In case of feature extraction on individual data blocks, modifications can be localized to blocks or to even minute level. In certain cases it might even be possible to correct the identified erroneous blocks [12]. It is however not so common to have the extra ability of correcting the erroneous blocks in addition to data authentication in the same scheme.

3.6   Two Phase Soft Authentication

Certain two phase data authentication techniques have been proposed in literature. The aim of the two phase schemes is to authenticate data in two phases or steps. Another indirect aim is to reduce the probability of false acceptance of data by basing the authentication decision on two steps instead of one. Typically the first step is focused on trying to identify error locations or erroneous blocks using some techniques, such as using error correcting codes. In the second phase, the errors are typically tried for correction.

Depending on the method under discussion, either one or two or both of the phases have to pass together for authentication to succeed. Sometimes, certain thresholds have to be crossed in each phase for the data to be declared authentic. It is totally dependent on the method of data authentication.

3.7   Watermarking-Based Noisy Data Authentication

Watermarking techniques for noise tolerant authentication are based on hiding a marker/identifier into the data (signal, image, voice, video, etc.). This marker is extracted by the receiver to identify the ownership of the signal. Digital watermarks are embedded in such a manner that only a valid and intended recipient should be able to extract it. This is typically ensured by making the positions of the bits of watermark a secret, e.g., through pseudorandom permutation with a shared key as a seed.

Typically the watermark permanently affects the signal in which the watermark is embedded by rendering the bit positions of the signal useless. In some digital watermarking techniques, the watermark can be used to reconstruct the effected bits. This is typically achieved by using forward error correction codes. Sometimes, the parity of the forward error correction codes is used either directly or indirectly as a watermark. This is useful in reconstructing those parts of the cover signal which are either destroyed by legitimate signal processing operations or by embedding the watermark.

3.8   Conclusion

Noise tolerant data authentication mechanisms are discussed in this chapter. These methods are tolerant to minor modifications in the data and/or its authentication tag. Such methods include generic noise tolerant data authentication algorithms, content-based noise tolerant algorithms using feature extraction, noise tolerant authentication algorithms using FEC codes, and digital watermarking techniques.

References

1. R.F. Graveman, K.E. Fu, Approximate message authentication codes, in *Proceedings of the 3rd Annual Fedlab Symposium on Advanced Telecommunications/Information Distribution*, vol. 1, College Park, Feb 1999 (Cambridge University Press, 2005)
2. D. Tonien, R. Safavi-Naini, P. Nickolas, Breaking and repairing an approximate message authentication scheme. Discret. Math. Algorithms Appl. **3**(3), 393–412 (2011)
3. L. Xie, G.R. Arce, R.F. Graveman, Approximate image message authentication codes. IEEE Trans. Inf. Theory **56**, 4922–4940 (2010)
4. C. Ruland, N. Zivic, Soft input decryption, 4th Turbo-code Conference, 6th Source and Channel Code Conference, VDE/IEEE, Munchen, 3–7 April 2006
5. C. Berrou, A. Glavieux, P. Thitimajshima, Near Shannon limit error correcting coding and decoding: Turbo codes, Proceeding of IEEE International Conference on Communication, vol. 2/3, pp. 1064–1070, Geneva, 1993
6. J. Hagenauer, *Soft Is Better than Hard, Communications, Coding and Cryptology* (Kluwer Verlag, Leiden, 1994)
7. D. Chase, A class of algorithms for decoding block codes with channel measurement information. IEEE Trans. Inf. Theory **18**, 170–182 (1972)
8. C.G. Boncelet Jr., The NTMAC for authentication of noisy messages. IEEE Trans. Inf. Forensics Secur. **1**, 1 (2006)
9. O. Ur-Rehman, N. Zivic, S. Amir Tabatabaei, C. Ruland, Error correcting and weighted noise tolerant message authentication codes, 5th International Conference on Signal Processing and Communication Systems, Hawaii, Dec 2011

10. O. Ur Rehman, N. Zivic, Noise tolerant image authentication with error localization and correction, 50th Annual Allerton Conference on Communication, Control, and Computing, pp. 2081–2087, Allerton, 1–5 Oct 2012
11. D.G. Lowe, Distinctive image features from scale invariant keypoints. Int. J. Comput. Vis. **50**(2), 91–110 (2004)
12. C. Ling, O. Ur-Rehman, W. Zhang, Semi-fragile watermarking scheme for H.264/AVC video content authentication based on manifold feature. KSII Trans. Internet Inf. Syst. **8**(12), 4568–4587 (2014)

Chapter 4
Digital Watermarking for Image Authentication

Abstract This chapter serves as an introduction to digital watermarking, and various aspects pertinent to digital watermarking techniques for data authentication are discussed. The chapter is focused on multimedia authentication in general and image authentication in particular. Requirements for watermarking techniques are listed. A classification of watermarking techniques is given followed by a security analysis. Watermarking techniques for image authentication are briefly discussed in the end.

4.1 Digital Watermarking

Digital watermarking is the process of hiding or covertly embedding a digital information in a digital signal. The embedded signal is called the digital watermark. The signal in which the watermark is embedded is called the host or carrier signal. The carrier signal can be of any type such as audio, video, image, and text. However, in this book, only images are considered as host or carrier signals. A watermark is embedded in the carrier signal such that only an intended receiver can extract it to prove the authenticity and integrity of the carrier signal. This is typically done using a pre-shared information, such as a secret key shared between the transmitter and the receiver. This shared key is used to identify the location of watermark, extract it, and use it to authenticate the carrier image. For a comprehensive study on digital watermarking, the reader might like to consult [1, 2].

Typically, a digital watermarking scheme has the following three main stages.

4.1.1 Watermark Generation

Watermark should be such that its content is unique to the image that it protects. It is therefore typically based on the unique features of the image to be protected. The image features are found such that they are unique to the image to be protected and

O. Ur-Rehman, N. Zivic, *Noise Tolerant Data Authentication for Wireless Communication*, Signals and Communication Technology,
https://doi.org/10.1007/978-3-319-78942-2_4

then a watermark is generated based on those features. The image features by themselves are usually not used directly as a watermark, rather some preprocessing is typically performed on the features to generate a watermark. The preprocessing includes pseudorandom permutations with the knowledge of a shared secret (such as a cryptographic key) or also error correcting codes. After watermark is generated, it is embedded in the host or carrier image.

4.1.2   Watermark Embedding

The watermark should be embedded in the cover image in such a way that the cover image itself or its features are not destroyed by the embedding process. It should also be very difficult for someone who is not an intended receiver to identify the existence of the watermark in the image or its location. The aim is to make it extremely difficult for an attacker to locate the watermark and subsequently to replace or destroy it.

4.1.3   Watermark Extraction

Only the intended recipients of the message containing the watermark should be able to extract it. Watermark extraction is the reverse process of watermark embedding. The same secret information (key) is needed to locate and extract the watermark from the cover image. Once the watermark is completely extracted and recovered, the carrier signal or image can be authenticated and tampering can be detected.

4.2   Requirements of Digital Watermarking

The generation and use of digital watermarks is based on many requirements such as:

4.2.1   Invisibility

A digital watermark should be invisible in the sense that the changes induced by embedding the watermark should not be visible to the human eye.

4.2.2 Tamper Detection

A digital watermark should be able to detect tampering in the image it protects. Normally, when an image is tampered, the watermark bits are also destroyed. Therefore, tamper detection is supported by watermarking. However, it should also be possible to detect the tampered locations in the protected image. One of the ways to achieve this is possible through the use of error correcting codes when generating a watermark.

4.2.3 Robustness

Legitimate image processing operations at the transmission, such as compression and quantization, and channel noise during transmission, such as over a wireless medium, can degrade the image quality and modify the image data. However, the content of the image might remain intact. Watermarking should be robust to the abovementioned intentional and unintentional modifications which keep the image content intact. However, if the content of the image is not preserved, then the watermarking should be able to detect it.

4.2.4 Capacity

One of the requirements of a digital watermarking scheme is its capacity. Capacity is the number of bits of a watermark that can be embedded in the carrier signal. For example, there are various methods for inserting the watermark bits in the image such that the distortion is not distinctively visible. If watermark bits are inserted in a transform domain vs. inserting the watermark bits in the least significant or most significant bits of the pixel values.

4.3 Classification of Digital Watermarking

Digital watermarking schemes are often classified on the basis of the following criteria.

4.3.1 Embedding

A digital watermark might be embedded either in spatial domain or frequency transform domain. Each method has its pros and cons.

One of the most well-known methods of embedding watermark in the spatial domain is to embed it in a chosen combination of the least significant bits (LSB). If a byte is used to represent the color intensity of an image pixel, then changing the LSB does not reflect in big visual distortions in the image. A human eye is typically insensitive to such distractions. Thus LSB is a good method to hide watermark. However, once the combination of LSBs used for watermark embedding is known, anyone can extract the watermark, or replace it.

Another method of hiding watermark is in the frequency transform domain, spreading the watermark content over the frequency spectrum and making it difficult to detect its existence. The most common frequency transform domain watermark hiding techniques use discrete Fourier transform (FFT), discrete cosine transform (DCT), and discrete wavelet transform (DWT). The first two approaches are related and are both Fourier transforms. DFT produces data which has both real and imaginary components. However, DWT has only real numbers. In both approaches, which are typically used for images, an image is divided into equal sized nonoverlapping blocks. The DFT or DCT is calculated on each block and the watermark bits are spread over the resultant blocks. DWT splits the given data into four frequency levels categorized as low low (LL), high low (HL), low high (LH), and high high (HH). Embedding the watermark in the low frequency part (LL) can damage the cover image to a higher extent as compared to the high frequency part (HH). Thus, a watermark is typically not embedded in the LL part, rather in the higher energy components such as the LH, HL, or the HH.

4.3.2   Robustness

Another criterion to classify watermarking schemes is based on the robustness of the scheme. This ultimately refers to the ability of watermark detection and extraction, even after various legitimate and illegitimate image processing operations. The legitimate operations include compression, quantization, storage, transmission noise, etc. Illegitimate image processing operations include tampering attacks such as object insertion, and object removal.

Watermarking schemes can be categorized into fragile and non-fragile watermarking based on the criteria of robustness:

- Fragile watermarking: In fragile watermarking, the watermarks are destroyed even if very minor manipulations occur in the data they protect. This makes fragile watermarking a good choice for integrity verification.
- Semi-fragile watermarking: Semi-fragile watermarking schemes are not as sensitive to modifications as fragile watermarking schemes. In semi-fragile watermarking, small changes in the data protected through the watermark are tolerable. Such small changes include the transmission noise, compression, etc.

4.3.3   Perceptibility

Watermarking schemes can also be classified on the basis of the perceptibility criterion. Based on these classifications, the digital watermarking scheme is either perceptible or non-perceptible. In perceptible schemes, the watermark is typically visible, such as the logo appearing on an image or video. This is a classical approach used to identify the ownership of the protected data. In a non-perceptible watermarking scheme, the watermark is not visible to a naked eye, however, it can be located using a secret (pre-shared) information. All of the watermarking schemes discussed in this book are invisible or non-perceptible.

4.3.4   Compression

Watermarking schemes can also be classified on the basis of the compression method they use. Text, images, audios, and videos are typically compressed at the source to reduce their size. This is helpful for storage and transmission. The compression methods are either lossy or lossless.

In lossless compression, the source data can be perfectly reconstructed from the compressed data. In lossy compression the original data cannot be perfectly reconstructed, hence named lossy. Those application areas where loss of data is not tolerable use lossless compression. However, if loss of data is tolerable, then lossy compression can be used. Examples of lossy compression standards are JPEG and JPEG2000. Watermarking can be integrated into source coding; however, it is typically done after the compression has been performed.

4.4   Conclusion

Digital watermarking techniques are introduced in this chapter. This serves as a foundation for the watermarking algorithms discussed in the following chapters. At first, the requirements for digital watermarking schemes are discussed. The criteria for classification of digital watermarking methods are briefly listed. Finally, the security issues concerning digital watermarking methods are discussed briefly. Digital watermarking methods are used later in other algorithms discussed in the following chapters.

References

1. C. Ling, O. Ur-Rehman, Watermarking for image authentication, in *Robust Image Authentication in the Presence of Noise*, ed. by N. Zivic (Springer, New York, 2015), pp. 43–74
2. L. Kumar Saini, V. Shrivastava, A survey of digital watermarking techniques and its applications. Int. J. Comput. Sci. Trends Technol. 2(3) (May-Jun 2014)

Chapter 5
Dual Watermarking

Abstract In this chapter, two dual watermarking techniques are discussed. Both of these methods are based on extracting features from an image and using the features to generate watermarks. The watermarking techniques discussed in this chapter are based on a combination of image features and forward error correction codes. It is shown how the ability of tolerance to modifications, location of modified positions and (partial) reconstruction of image is achieved through the discussed method. Simulation results are presented for different channels and using different codes.

5.1 Watermarking in Multiple Domains

Digital watermarking has many applications such as owner identification, copyright protection, and content authentication. A digital watermark should have certain qualities, making it suitable for embedding digital data with secret information. These include the following among many other:

- A digital watermark should be unique, so that it can later be used for authentication.
- A watermark should be complex, making it difficult for an attacker to extract and damage or replace it.
- A watermark should be such that extracting it damages the cover object.

Watermarks are typically generated using image features. The feature vector (f) of the cover image is extracted using a feature extraction function ExtractFeatures($\cdot$),

$$f = \text{ExtractFeature}(\text{Image}) \tag{5.1}$$

The features uniquely identify the cover image. This means that two (different) images will have different features but if the images have the same content, then they will have the same features, even if they are represented by different data.

© Springer International Publishing AG, part of Springer Nature 2018   39
O. Ur-Rehman, N. Zivic, *Noise Tolerant Data Authentication for Wireless Communication*, Signals and Communication Technology,
https://doi.org/10.1007/978-3-319-78942-2_5

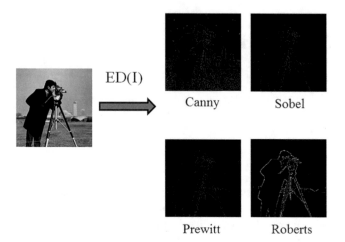

Fig. 5.1 Standard edge detectors

A watermark w is generated on the image using a watermark generation function which uses the feature vector, f, and a secret key, k, as parameters,

$$w = \text{GenerateWatermark}\left(f,k\right)$$ (5.2)

The watermark, w, can be embedded in the cover image either directly into the image data (i.e., in spatial domain) or by transforming the image to be protected into another domain (e.g., frequency domain). The receiver can extract the watermark from the cover image, using the key k, to verify its authenticity. The dual watermarking approaches presented in this chapter are based on computing two watermarks, one in the spatial domain and the other one in the frequency transform domain. In the first approach, the watermark in the spatial domain is based on the edge data which is extracted from the image to be protected. The watermark itself is protected with AMAC tag. The second watermark, which complements the first watermark, is based on the frequency domain features, such as discrete cosine transform (DCT). These features are then protected using systematic Reed-Solomon (RS) codes, which help in error localization and correction.

5.1.1   Edge Detection

Edge detection methods detect discontinuities in the image brightness by identifying sharp changes in the brightness levels. Normally two different images contain different objects which will have different edges or at least they will have different positions. Standard edge detection techniques used in image processing, include Canny [1], Sobel [2], Prewitt [3], and Roberts edge detectors [4]. The edges detected by these function for a sample cameraman image are shown in Fig. 5.1.

5.1.2 *Discrete Cosine Transform*

Discrete cosine transform (DCT) of an image makes spectral analysis of the image and orders the spectral regions from high energy to low energy. DCT can be applied to the whole image to obtain global features. Another approach would be to split the image into $N \times N$ blocks, where N is a positive integer, and apply DCT to each block such as

$$F(u,v) = c(u)c(v) \sum_{N-1}^{i=0} \sum_{N-1}^{j=0} f(i,j) \cos\left[\frac{\pi(2i+1)}{2N}u\right] \cos\left[\frac{\pi(2j+1)}{2N}v\right]$$

$$0 \le u \le N-1, 0 \le v \le N-1$$

$$c(u) = \begin{cases} \sqrt{\frac{1}{N}}, u = 0 \\ \sqrt{\frac{2}{N}}, u \ne 0 \end{cases}, c(v) = \begin{cases} \sqrt{\frac{1}{N}}, v = 0 \\ \sqrt{\frac{2}{N}}, v \ne 0 \end{cases}$$

(5.3)

Inverse DCT is used for reverse conversion. The inverse DCT transform for each block shown above is computed as

$$f(i,j) = \sum_{N-1}^{u=0} \sum_{N-1}^{v=0} c(u)c(v) F(u,v) \cos\left[\frac{\pi(2i+1)}{2N}u\right] \cos\left[\frac{\pi(2j+1)}{2N}v\right]$$

$$c(u) = \begin{cases} \sqrt{\frac{1}{N}}, u = 0 \\ \sqrt{\frac{2}{N}}, u \ne 0 \end{cases}, c(v) = \begin{cases} \sqrt{\frac{1}{N}}, v = 0 \\ \sqrt{\frac{2}{N}}, v \ne 0 \end{cases}$$

(5.4)

Figure 5.2 shows the image of an eye region. Consider a block of the image focused on the retina of the eye and highlighted using a square. The square is shown as pixel intensities along with the DCT of the block.

5.1.3 *Dual Watermark Generation and Embedding at the Source*

Watermark₁ In order to generate the first watermark, watermark$_1$, edges of an image are extracted using a standard edge extractor, such as the Canny edge detector. Edged image has the same dimensions as the image whose edges are extracted. When represented in binary form, the pixels indicating an edge are represented by 1 and the pixels representing the absence of an edge are represented by 0. This binary edge data matrix is reshaped into $L \times R \times S$ bits for AMAC tag generation, where L

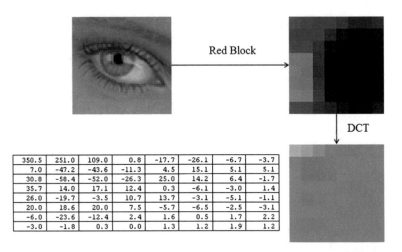

Fig. 5.2 Discrete cosine transform

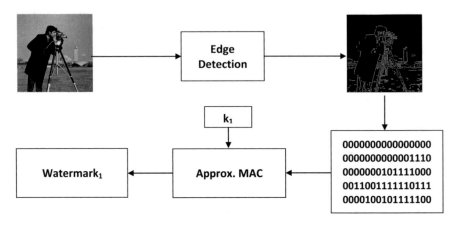

Fig. 5.3 Watermark₁ generation at the source using AMAC

is the length of AMAC in bits and R and S are positive integers as described in Chap. 3. The secret key k_1 is pre-shared between the transmitter and the receiver. The first watermark (denoted as watermark₁) is based on AMAC.

The process of generating watermark₁ is shown in Fig. 5.3.

Watermark₂ In the proposed approach, frequency domain watermark based on DCT is used for tamper detection and localization. This watermark (denoted as watermark₂) complements the watermark₁. Watermatk₁ and watermark₂ are combined together to produce the final watermark. An image is divided into blocks of 8×8 pixels. The DCT of each block is computed and the DC elements are chosen for watermark generation. Each DC element is 16-bit long. Reed-Solomon code

$dc_{1,1}$	$dc_{1,2}$	$dc_{1,3}$	....	$dc_{1,N}$	RS-Parity$_{r1}$
$dc_{2,1}$	$dc_{2,2}$	$dc_{2,3}$	....	$dc_{2,N}$	RS-Parity$_{r2}$
$dc_{3,1}$	$dc_{3,2}$	$dc_{3,3}$	....	$dc_{3,N}$	RS-Parity$_{r3}$
		....		....	
		....		....	
$dc_{N,1}$	$dc_{N,2}$	$dc_{N,3}$	....	$dc_{N,N}$	RS-Parity$_{rN}$
RS-Parity$_{c1}$	RS-Parity$_{c2}$	RS-Parity$_{c3}$	⋮	RS-Parity$_{cN}$	

Fig. 5.4 Protecting the DC elements with RS codes

RS(255, 239) is shortened according to the image resolution. For example, for a 128×128 pixel image with 8×8 blocks, there are 16 blocks in a row and therefore 16 DC elements should be protected row-wise. An RS(48, 32) code, shortened from RS(255, 239) provides an error correction capability of eight symbol errors. Each 16-bit DC element is split into two 8-bit RS symbols, and therefore up to four erroneous DC elements can be corrected by the eight symbol error correction capability. However, by using an erasure correcting RS decoder, this is extended to eight DC elements.

The DC elements of all the blocks are arranged in rows and RS encoding is done row-wise as well as column-wise as shown in Fig. 5.4. Sixteen RS parity symbols (RS-Parity$_{ri}$) are obtained for each row, r_i, where $1 \leq i \leq N$ and N is the number of blocks in a column (and also in a row, in case of a square image). The process of systematic encoding using RS codes is repeated column-wise, where the parity symbols RS-Parity$_{ci}$ are obtained for each column c_i. This approach provides a double protection to each DC element and helps in locating the potentially erroneous DC elements as well as correcting the marked erroneous DC elements.

If the RS parities are used directly as watermark$_2$, a probability exists that an attacker will be able to forge the DC components by extracting the original ones, modifying the image and replacing the original DC elements by the newly obtained ones for the forged image. The RS parities are XORed with a pseudorandom bit stream based on a secret key k_2 known to both the transmitter and the receiver. For each bit x_i of the RS parity, two pseudorandom bits y_i, z_i are produced. Each bit w_i of the watermark$_2$ is obtained as given in (3) and the process of generating watermark$_2$ is shown in Fig. 5.4. An extra pseudorandom bit (y_i) is used in (3) as

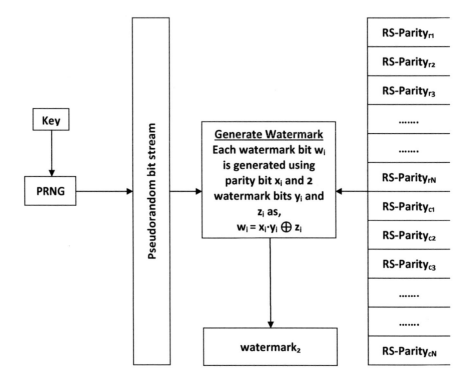

Fig. 5.5 Protecting the DC elements with RS codes

compared to the original approach for AMAC calculation, which would calculate an XOR of x_i and z_i.

The process of generating watermark$_2$ is shown in Fig. 5.5.

5.1.4   Watermark Embedding

The final watermark is generated by interleaving the bits of watermark$_1$ and watermark$_2$. This watermark is embedded in the AC coefficients of each DCT block of the cover image. The DC element and the AC elements of the first three diagonals are not used for watermark embedding to avoid image degradation. The least significant bits (LSBs) of the next few AC components are chosen for watermark embedding. If an AMAC tag of 256 bits is chosen and there are 16 × 16 blocks in an image, then each watermark$_1$ bit is distributed to one image block. In other cases, a different mapping would be obtained. The bits from watermark$_2$ are also embedded based on the image resolution. For 16 × 16 blocks in an image, there are 16 RS codewords row-wise and 16 RS codewords column-wise with 16 symbol parities each. RS parity for one codeword requires 16 × 8 = 128 bits. For the 32 RS codewords, the size of watermark$_2$ is $128 \times 32 = 4096$ bits. When they are distributed over the 128 image blocks, each image block gets 32 bits of watermark$_2$. Thus, in the example described above, a total of 33 bits of watermark (watermark$_1$ plus watermark$_2$) are assigned to each block. These are inserted into the two LSBs of the 17 AC components for each block, following the first 3 minor diagonals of AC components. This is done in order to minimize the visual degradation in the quality of the processed image.

5.1.5   Watermark Extraction

The edge data for the received image I' is generated at the receiver. An AMAC tag (AMAC'') on this data is recalculated using the same procedure as used at the source.

If AMAC'' is the same as the received AMAC', it is inferred that the edges of the transmitted and the received images are either the same or very similar to each other, due to the tolerance of the AMAC algorithm. For watermark$_2$ verification, the DCT of each block of the image (I') is recomputed block wise and the DC components of all the DCT blocks are arranged into rows and columns of $N \times N$, where N is the number of blocks. RS-parities' (from the watermark$_2$') are appended (row-wise and column-wise) to these DC elements to make the RS codewords. RS decoding is performed row-wise and column-wise. The decoding failure in RS codes is used to identify the error locations by using the RS error locator polynomial for each row/column of DC elements. The DC elements found to be erroneous are marked as suspicious and therefore the corresponding block for each suspicious DC element is also marked as suspicious. It is not necessary that these blocks are in error or are forged because of the possibility of a decoding error. Therefore, they are called suspicious and not erroneous blocks, in this work. The whole image is considered authentic when the verification using both of the watermarks (watermark$_1$ and watermark$_2$) succeed together.

5.1.6   Error Localization and Correction Using Extracted Watermark

The image blocks corresponding to the marked suspicious DC elements are used to localize the erroneous positions in the image. If a DC element is correctable using the RS decoder, the corresponding image block is marked as authentic. The reason for this is that the DC element represents the most energy in a block and therefore the image block is declared authentic if the DC element is authentic (or correct). If errors in the DC elements are not correctable, the erroneous blocks are only identified (marked suspicious) and the application can decide if it wants to discard the whole image or accept it partwise.

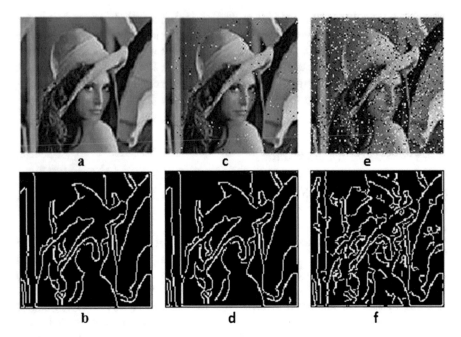

Fig. 5.6 Authentication in the presence of "Salt & Pepper" noise

5.2   Simulation Results and Analysis

5.2.1   Simulation Setup and Results

Grayscale images of 128×128 pixels are used in simulations. Each image is divided into blocks of 8×8 pixels. The length of AMAC is chosen to be 256 bits. Systematic RS(255, 239) codes shortened to RS(48, 32) are used. Each 16-bit DC element is represented by two RS symbols. Edge detection is performed using Canny edge detector.

Figure 5.6a shows the original Lena image followed by the edged image shown in Fig. 5.6b. Figure 5.6c shows the Lena image in the presence of "Salt & Pepper" noise of magnitude 0.01. The edged image corresponding to Fig. 5.6c is shown in Fig. 5.6d. Since the noise level is low, the edges are not much distorted and the authentication test based on watermark$_1$ succeeds due to the tolerance of AMAC. Because of the low noise (magnitude 0.01) combined with the error correction capability of RS codes, the same is valid in the frequency domain based watermark$_2$ and the image is declared authentic by the proposed dual watermarking approach.

Figure 5.6e shows the modified Lena image in the presence of "Salt & Pepper" noise of magnitude 0.1. Though the noise level is higher as compared to Fig. 5.6c, the edge distortion is still within the threshold of AMAC, and therefore the watermark$_1$ based authentication succeeds; however, it fails the frequency domain

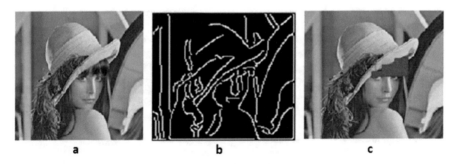

Fig. 5.7 (**a**) Forged Lena image (**b**) edge features (**c**) modifications detected using watermark$_2$

test based on watermark$_2$ due to a large variation in the DC elements which RS codes fails to correct.

A forged Lena image is shown in Fig. 5.7a, with extra hair on the forehead. Figure 5.7b shows the edges which pass the authentication check of watermark$_1$. However, the DC elements in the forged region are changed such that the RS based watermark$_2$ is able to locate the forged area but not able to correct it in the given example. The reason is that the number of modifications is beyond the error correction capability of RS codes. This shows that RS codes might not always be able to correct the modifications but they are detected and the forged area is located.

5.2.2   Security Analysis

In addition to the general security analysis of a hash-based image authentication scheme, two main potential attacks are worth considering. The first attack is a key recovery attack which discloses the secret key of the scheme using a sufficient number of authenticated image-hash pairs. The attacker can then use the recovered secret key to generate a hash of her own image to deceive the receiver with an impersonated image-tag pair. The second attack is a substitution attack, where the attacker tries to substitute a valid image-tag with another authentic image-tag pair. The attack is successful when the substituted image is perceptually different than the original valid image, while the difference between their tags is below the threshold value. This attack can be used as a basis for different adversary models. Therefore, the success rate of the attacker in mounting this attack is estimated here.

5.3   Variant Scheme Based on Discrete Wavelet Transform

5.3.1   Discrete Wavelet Transform

Discrete wavelet transform (DWT) is used to decompose an image hierarchically. Wavelet transform decomposes the image into band limited, low and high frequency components, which can be reassembled to reconstruct the original image. A DWT

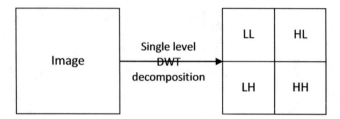

Fig. 5.8 Single level DWT decomposition

Fig. 5.9 Single level DWT
of Lena image

operation decomposes an image into four components represented as LL, LH, HL, and HH and as shown in Fig. 5.8, where L represents applying a low pass operation and H represents applying a high pass operation. Here LL is the low resolution approximation image and it closely resembles the original image. The other sub-bands, LH, HL, and HH represent other details such as edges. An example DWT of the Lena image is shown in Fig. 5.9.

5.3.2 Watermark Generation

Watermark is generated as follows. DWT is computed on the source image. The LL sub-band of the DWT is taken and passed through the AMAC algorithm. As a result, the AMAC tag is taken as the watermark of the image. If there are minor changes in the image, the LL sub-band will not change much and thus the AMAC will remain

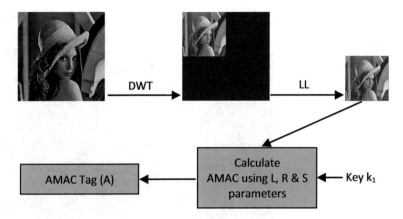

Fig. 5.10 Watermark generation for Lena Image

the same. For changes beyond a threshold, such as in case of forgery attacks, e.g., object insertion or object removal, the AMAC will change. The threshold is adjustable as discussed in the section on AMAC. The length of AMAC tag is chosen to be 256 bits. The watermark generation process for an example of Lena image is shown in Fig. 5.10.

5.3.3 Watermark Embedding

The watermark is embedded in the cover image. In the described method, the watermark is self-embedded in the source image. The source image is split into 8 × 8 pixel nonoverlapping blocks. The length of the AMAC tag is 256 bits. The AMAC tag is split into 32 sub-AMACs of 8 bits each. One sub-AMAC is taken at a time and inserted into the LSBs of the first 8 pixel values of the next image block which is obtained using a secret permutation. The next image block is chosen (pseudo) randomly, using a secret key, k_2, as the seed value. Thus, the AMAC tag is scrambled in the LSB of image blocks. This makes it hard for an attacker to extract and replace the watermark without the knowledge of the secret key. The watermarked image is the output of the watermark embedding process.

5.3.4 Watermark Extraction

When the authenticity of the image has to be proven, the watermark is extracted back from the image. The watermarked image is taken and split into 8 × 8 pixel nonoverlapping blocks. The LSBs of the first 8 pixel values of each next block is taken and appended with the previous to obtain the watermark (the AMAC tag).

Fig. 5.11 Authentication
in the presence of COFDM
with rate 1/2 LDPC code

The next block is chosen again using the pseudorandom permutation based on the shared secret key, k_2, to obtain the same sequence as obtained in the watermark embedding procedure.

5.3.5 Image Authentication

An image is verified by comparing the extracted watermark with the (re)computed watermark. Since the watermark is embedded in the spatial domain, it distorts a part of the cover image, which in this case is the source image as well. However, since AMAC is tolerant to modifications below the chosen threshold, the authentication will succeed even if there are minor deviations from the original.

5.3.6 Simulation Results for COFDM

Results obtained in the presence of COFDM transmission are shown in Fig. 5.11. COFDM transmission is considered in the presence of regular LDPC codes of rate 1/2 as the forward error correction codes. As all, or most, of the errors are corrected by the LDPC decoder, the image is declared authentic by the proposed algorithm.

5.4 Conclusion

In this chapter, two dual watermarking techniques for noisy data authentication are discussed. They are based on extracting image features and protecting them with forward correction codes to generate the watermarks. In one of the methods, edge detection methods are used for feature extraction, whereas in the second method DWT-based features are generated.

References

1. J. Canny, A computational approach to edge detection. IEEE Trans. Pattern Anal. Mach. Intell. **PAMI-8**(6), 679 (1986)
2. R. Gonzalez, R. Woods, *Digital Image Processing*. Boston, US (Addison Wesley, 1992), pp. 414–428
3. J.M.S. Prewitt, *Object Enhancement and Extraction, Picture processing and Psychopictorics*. Boston, US (Academic Press, 1970)
4. L.G. Roberts, Machine perception of three-dimensional solids, Massachusets Institute of Technology, Lincoln Laboratory, May 1963

Chapter 6
Enhanced Decoding Based on Authenticating the Region of Interests

Abstract In this chapter, a method for image authentication is described which also supports enhanced decoding. The method works in two steps at the receiver. In the first step only important parts of image are authenticated. The important parts are pre-marked regions of importance in the image such as the retina of an eye, number plate of a car. When the first phase succeeds, the second phase follows. In the second step, the successfully authenticated pre-marked important segments are used for improved decoding of the image. For this purpose, additional information is used from the channel as a side information. It is shown that the results produced by the two-phase method are superior both for authentication as well as for error correction.

6.1 Enhanced Decoder

The method described in this chapter uses the soft input decryption (SID) [1] technique discussed earlier for image authentication. The SID technique is combined with the turbo decoder [2] for enhanced decoding at the same time. The bit reliabilities produced by the turbo decoder are used in the first phase by SID for soft input authentication. If the first phase is successful, the second phase is executed, otherwise the image is declared unauthentic and there is no attempt for image reconstruction. The second phase uses the corrected and authenticated data bits from the first phase. These bits are called "known bits" and they are used a feedback to turbo decoder in a way that the image correction is improved beyond the results achievable by the turbo decoder alone.

The concept of using data known to both the transmitter and receiver to enhance the results of channel decoder was demonstrated in [3], where "dummy bits" were embedded in the transmitted data at pre-chosen locations. The dummy bits can be chosen in the simplest case to have values of all zeros or all ones. The positions of dummy bits are known at the transmitter and receiver. Since these bits are "dummy," their values are not important, rather only their positions. In [3], the dummy bits are

© Springer International Publishing AG, part of Springer Nature 2018   53
O. Ur-Rehman, N. Zivic, *Noise Tolerant Data Authentication for Wireless Communication*, Signals and Communication Technology,
https://doi.org/10.1007/978-3-319-78942-2_6

inserted in a zigzag manner in the data and are spread in this manner over the entire data. The receiver sets the reliability values of the dummy bits to $\pm\infty$. A reliability value of $\pm\infty$ means that the value of the bit is known with a highest (absolute) certainty. This is due to the bits which are already known at the receiver; therefore, the reliability of the bit value is 100%. The decoding of the entire data received by the receiver is performed using maximum a posteriori (MAP) decoder [4]. It was shown in [3] that an improvement in the decoding results is obtained by using dummy bits.

The transmitter and receiver for the two-phase method are described next.

6.2   Transmitter Side

At the transmitter, important parts are chosen in the image to be authenticated. Each of these important parts is named as a region of interest (RoI). A RoI can be a part of image such that it is unique to that particular image, e.g., iris of an eye, a finger-print, or a number plate on car; however, the uniqueness is not necessary. A RoI is extracted from the image I and a MAC tag, t, is computed over it using a standard MAC algorithm. A standard MAC algorithm uses a secret key, k, shared with the receiver. The computed MAC tag is appended to the image and both the image and its tag are encoded using turbo codes. The encoded data is either stored on a storage medium or transmitted over a potentially noisy communication channel. The RoIs can be identified statically at the transmitter as well as at the receiver, or they can be detected automatically using advanced techniques for object detection, such as those described in [5, 6], but these methods are not discussed in this writing. The receiver can easily identify the same RoIs as identified by the transmitter, if the transmitter appended the coordinates of the RoIs to the image before encoding and has transmitted these inside the encoded data.

The transmitter side is shown in Fig. 6.1.

6.3   Receiver Side

The receiver operates in two steps on the data received over a potentially noisy channel or storage medium. The received image and its authentication tag (MAC tag) might be erroneous. These are passed through the channel decoder. The results shown later on are based on a decoder for turbo codes but in general any decoder which output reliability values of the decoded bits can be used. The decoded data might still be erroneous due to some non-corrected errors after decoding (as the number of errors might have been more than the error correction capability of the used code). The decoded image and MAC tag pair is represented by (I', t') to differentiate it from the transmitted pair (I, t). The decoder also outputs the reliability values or the LLRs corresponding to each decoded bit [7].

Fig. 6.1 Transmitter

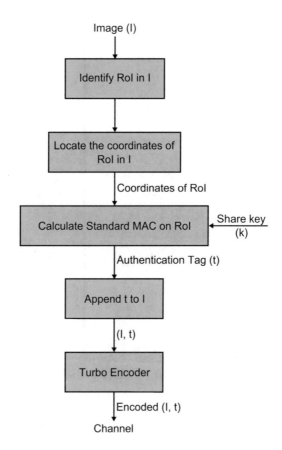

The general idea of the receiver is shown in Figs. 6.2 and 6.3.

The receiver performs the enhanced decoding and authentication in two steps as follows.

6.3.1   Step 1

RoI (assuming that there is only one) is extracted by the receiver from the decoded data and denoted as RoI′. The receiver performs soft input decryption on the RoI extracted from the decoded data. For SID, a new MAC tag t'' is calculated on the RoI′ using the shared key and compared with the received and decoded tag t'. If the tag verification is unsuccessful, SID iterations are performed using the RoI′ together with t', t'' and the LLRs for the RoI′ and t' [9]. The LLRs are typically obtained from a channel decoder, such as the decoder for convolutional codes or turbo codes. However, when a channel decoder is not used, the LLR values can also be obtained

Fig. 6.2 Receiver side

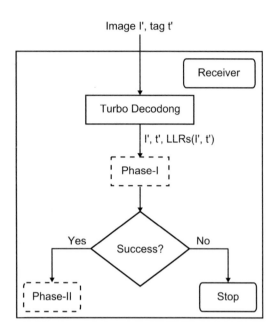

directly from the channel observations. The total number of iterations N_r of SID is dependent on the number of least reliable bits, m, used in the SID algorithm and is denoted by [7],

$$N_r = 2^m \tag{6.1}$$

If the SID algorithm finds a valid RoI′ and t′ pair, the SID algorithm is successfully finished. Thus the RoI and t can be used as "known data" at the receiver for the second phase. This is similar to the concept of "dummy bits" as explained earlier, where known bits are inserted at predefined positions by the transmitter and the receiver can set their reliabilities to $\pm\infty$. Therefore, the LLRs of the bits corresponding to the successfully authenticated RoI are set to $\pm\infty$. This is based on the fact that the RoI data is authentic and sent by the transmitter with whom the receiver shares the secret key k. These known data bits are used to improve the decoding results of the turbo decoder in the next step.

If SID is not successful, the data is declared unauthentic and further processing stops.

6.3.2   Step 2

The Step 2 is focused on enhanced data decoding as opposed to Step 1, which was focused on RoI authentication [7]. The already authenticated RoI and its tag t are used together with their corresponding LLRs. These LLR values are set to the

Fig. 6.3 Receiver side

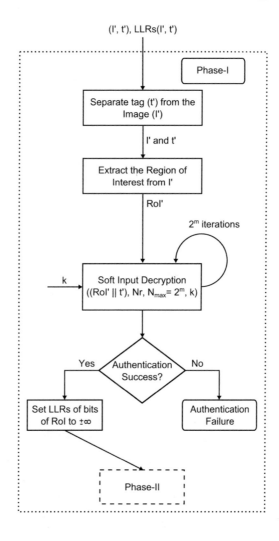

maximum possible theoretical value in Step 1. The objective here is to improve the decoding results of the used channel decoder (e.g., the turbo decoder). The RoI, the corresponding tag t and the LLRs are used as feedback to the turbo decoder with the aim of enhanced decoding based on the corrected and authenticated RoI. The RoI and its tag bits are spread over the entire image I' using a pseudorandom permutation. The turbo decoder is executed again with the feedback (maximal absolute values of LLRs). After decoding, the inverse permutation is performed to bring the bits of RoI back to their original positions in the decoded image.

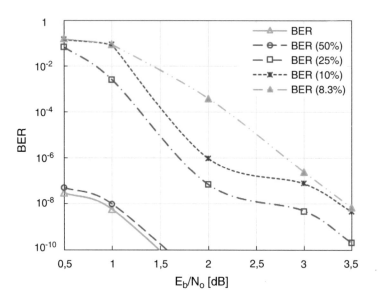

Fig. 6.4 Bit error rate

6.4   Simulation Results

Simulation results are given in this section for the two-step authentication and decoding method. In order to obtain the simulation results, Lena image of various resolutions is considered. The channel decoder is a turbo decoder of rate-1/3 whereas additive white Gaussian noise (AWGN) is considered. Simulation results are based on a MAC tag length of 160 bits; however, it is recommended to use a larger MAC tag, such as 256 or 512 bits, for a higher level of security. The maximum number of iterations for SID is chosen to be $N_r = 2^8 = 256$. As discussed earlier, the RoI is statically identified in this work.

The known bits are used for improving the bit error rate (BER) as also supported by the simulation results in this section. Simulation results for different number of known bits are presented in Fig. 6.4. It shows the BER vs SNR results when 12 decoding iterations of turbo codes are chosen and the RoI authentication succeeds. The curves show the different number of known bits. The curve BER shows the bit error rate in the absence of known bits. The curve BER (50%) shows the case when 50% of the bits are known at the receiver, BER (25%) depicts the case when 25% of the bits are known at the receiver, and so on. Among the shown curves, the best BER curve is obtained for the case when 50% of the bits are known at the receiver and worst is for the case when 8.3% of the bits are known at the receiver.

Figure 6.5 shows the Lena image in three different resolutions. For each resolution, phase-II of the proposed method is applied when 50%, 25%, and 10% of the

Resolu- tion	Original image	50% bits are known at receiver	25% bits are known at receiver	10% bits are known at receiver
32 × 32				
64 × 64				
128 × 128				

Fig. 6.5 Enhanced error correction using known data with turbo codes of rate-1/3 and six iterations

bits are known. The results are measured in the presence of AWGN channel at an SNR of 2.5 dB and turbo codes with three iterations. However, the improvement in the presence of known bits remains in the similar proportion in case of other SNR values.

6.5 Conclusion

A two-step method for image authentication followed by enhanced decoding based on the feedback of authenticated RoIs is presented. The described method works in two steps at the receiver. The first step authenticates only important parts of image (RoI), whereas in the second step, the successfully authenticated image segments are used for improved decoding of the entire image. For this purpose, additional information is used from the channel decoder as side information. The simulation results show that the results produced by the two-phase method are superior both for authentication as well as for error correction.

References

1. C. Ruland, N. Zivic, Soft Input Decryption, *6th International ITG Conference on Source and Channel Coding*, Apr. 3–7, 2006, Munich, Germany
2. C. Berrou, A. Glavieux, P. Thitimajshima, Near Shannon Limit Error Correcting Coding and Decoding: Turbo-Codes. 1. *IEEE International Conference on Communications (ICC '93)*, May 23–26, 1993, Geneva, Switzerland

3. C. Ruland, N. Živic, Feedback in Joint Coding and Cryptography. *7th International ITG / IEEE Conference on Source and Channel Coding*, January 2008, Ulm, Germany
4. L. Bahl, J. Cocke, F. Jelinek, J. Raviv, Optimum decoding of linear codes for minimizing symbol error rate. IEEE Trans. Inf. Theory **IT-20**, 284–287 (1974)
5. Y T. Zheng, M. Zhao, Y. Song, H. Adam, U. Buddemeier, A. Bissacco, F. Brucher, T. Chua, Tour the World: Building a Web-Scale Landmark Recognition Engine. *2009 IEEE Conference on Computer Vision and Pattern Recognition*, Miami, FL, 2009, pp. 1085–1092
6. C.A.R. Behaine, J. Scharcanski, Enhancing the performance of active shape models in face recognition applications. IEEE Trans. Instrum. Meas. **61**(8), 2330–2333 (2012)
7. O. Ur-Rehman, N. Zivic, Two-phase method for image authentication and enhanced decoding. IEEE Access **5**(1), 12158–12167 (2017)

Chapter 7
Authentication with Block Level Error Localization

Abstract In this chapter, two noise tolerant data authentication algorithms are presented which are able to localize errors to the block level. As an application, image data is considered, whereby the image is divided into equal sized blocks. The proposed algorithms are able to tolerate minor modifications in the image, and at the same time, they are able to identify the major errors or forgeries. The modifications can also be localized by the proposed algorithms to the block level. It is also possible to correct the modifications to some extent. If the modifications are below a certain predefined threshold, they will be corrected. If all of the modifications are corrected, the image is declared authentic. If some modifications are not correctable, it is left for the application using the algorithms to decide if these modifications can be tolerated and the image can be partly accepted. This makes sense when retransmissions are not possible over a communication channel, e.g. in case of satellite transmission. Simulation results are presented to show the effectiveness of the scheme.

7.1 Authentication with Error Localization and Correction

Multimedia data authentication together with error localization and correction has been studied before. Each method has its own pros and cons. The general idea is to combine message authentication codes or digital watermarking together with forward error correction codes. The combination is used to identify, localize and correct modifications in the multimedia data. However, it is very important that an effective scheme is able to differentiate unintentional modifications from forgeries.

In [1], a watermarking method for authenticating as well as restoring lost or tampered information in images or videos is proposed. The proposed method is able to localize errors and also to retrieve the original content by restoring the manipulated information. The watermark is generated using a halftoning technique, and a quantization index modulation technique is used to embed this watermark into the protected picture. An inverse halftoning algorithm is used to restore a multilevel

© Springer International Publishing AG, part of Springer Nature 2018   61
O. Ur-Rehman, N. Zivic, *Noise Tolerant Data Authentication for Wireless Communication*, Signals and Communication Technology,
https://doi.org/10.1007/978-3-319-78942-2_7

approximation of the original content. Another watermarking method is proposed in [2], which can localize the tampering regions in fine granularity. The proposed method encodes the block hash with a systematic error-correcting code. The parity symbols are embedded into the blocks. During verification, the hash of each block is recovered with the embedded parity symbols if the number of tampered blocks is within a threshold value. In [3], a fragile watermarking method for error detection in video streams is proposed. The watermark is embedded into the quantized DCT coefficients of the video blocks. If the block is tampered, errors exist in the bit stream associated with the block, and the bit stream errors can accurately be located at the block level. In [4] a method for approximate authentication and correction of images is proposed. The proposed method is constructed using two approximate message authentication code frameworks. The frameworks are based on standard cryptographic primitives and error-correcting codes. The method tolerates acceptable modifications below a predefined threshold. The malicious modifications are detectable, and the modifications below a certain threshold are corrected. For a good study of robust image authentication methods in the presence of noise, the interested readers should refer to [5].

7.2   Building Blocks of the Authentication Method

The building blocks of the authentication method discussed in this chapter are discussed here briefly.

The authentication schemes, with error localization and possibly also correction, discussed in this chapter are based on NTMAC and WNTMAC which were introduced in Chap. 3. As a quick reminder, NTMAC is based on standard MAC algorithms and authenticates data using a shared secret key. In NTMAC, only a part of the MAC, called sub-MAC, is used instead of the whole MAC, and a certain level of tolerance to authentication is achieved. WNTMAC is based on NTMAC with the additional feature of using weights for marking important parts of the data to be authenticated and thus processing the relatively important parts differently than the parts of data which are relatively less important.

The features of image considered for content protection are based on the discrete cosine transform (DCT) . Both of the algorithms discussed in this chapter protect the DC coefficients of the blocks of an image using NTMAC and WNTMAC algorithms. The length of sub-MAC chosen for protections of DC elements is different in both the algorithms. In the error-correcting NTMAC algorithm, only the DC elements are protected using a MAC (sub-MAC), and therefore the length of sub-MACs is the same for all blocks of the image. In error-correcting WNTMAC, not only the DC elements but also the AC elements are protected using WNTMAC. The DC elements are labelled as more important than the AC elements, and the lengths of sub-MACs are adjusted accordingly. The sub-MACs used for DC elements are given a higher importance as compared to the sub-MACs for the AC elements. Thus DC elements are protected more as discussed later in the chapter.

7.3   General Assumptions

In the proposed algorithms, it is assumed that an image needs to be authenticated not only in the presence of unintentional modifications such as channel noise but also forgery attacks. Let the resolution of the image be N × N pixels, where N is a positive integer. If the image is not square, it can be preprocessed and padded with all zero or all one bits to make it a square image of resolution N × N pixels.

Let the image be divided into blocks of m × m pixels, such that m I N, where m is a small integer which is typically chosen to have a value of 8. DCT is calculated on each block. Let's also assume that the number of blocks in a row, as well as a column, is B due to the assumption of a square image. Another assumption is that the input image is transformed into greyscale before an authentication tag is computed on it. The original image may be transmitted by itself. The greyscale version is used only for the computation of the authentication tag. The same preprocessing is repeated at the receiver side on the received image.

7.4   Error Localizing and Correcting NTMAC (ELC-NTMAC)

The algorithm discussed in this section is able to protect data through NTMAC. It is tolerant to minor errors, which are tolerable through NTMAC and/or further correctable using the additionally employed method of soft input decryption. The algorithm is able to identify modifications and to locate the potentially modified blocks in the image. By controlling the granularity of the block size, error location can be controlled to a fine-grained level.

The image is split into blocks and a DCT is calculated on each block. The resultant DC elements, against each block, are arranged in a matrix. An NTMAC tag is calculated for each row of the DC elements in the matrix of DC elements. As mentioned earlier, a full MAC tag is calculated for each DC element, but only a small part of it is retained, called the sub-MAC. This could be the most or least significant s-bits of the MAC tag.

The sub-MACs calculated for each row of the DC elements are appended together to form the NTMAC tag for that row of the DC matrix. There are B rows in the matrix, and therefore B NTMAC tags are calculated in total. Now the same process of calculating NTMAC tags is repeated for all the columns, giving another B NTMAC tags. Thus there are 2•B NTMAC tags in total which are transmitted together with the original image. The NTMAC tags will help in error localization to the block level.

The receiver receives the original image as well as the MAC tag, i.e. the 2•B NTMAC tags. The receiver recalculates the 2•B NTMAC tags using the same procedure as carried out at the transmitter. It then compares the received NTMAC, denoted as NTMAC′, with the recalculated NTMAC, denoted as NTMAC″. This essentially means that the corresponding sub-MACs against each block are compared. The comparison uses SID technique for authentication and correction of the

DC element together with the sub-MAC. If the result of tag authentication is successful, the block is marked as authentic. Note that this will be the case even in the presence of minor errors, which are corrected by SID.

If a DC element is accepted as authentic, it means that the corresponding message blocks are also accepted as authentic. If that is not the case, then the block is marked as suspicious. The suspicious blocks are potentially erroneous. It is therefore possible to perform correction to some extent and mark the block if correction is not possible. It is up to the application using the ELC-NTMAC algorithm to decide what to do with the suspicious blocks. They can be retained and recovered using techniques from image processing such as replacing them by the average pixel value of the neighbouring blocks, etc.

7.5 Error Localizing and Correcting WNTMAC (ELC-WNTMAC)

This algorithm is a slight modification of the ELC-NTMAC algorithm discussed in the previous section. In this algorithm WNTMAC is used as opposed to NTMAC. However, an additional difference is that after taking a DCT, not only the DC element but also a few AC elements are protected with the WNTMAC. However, a higher weight is assigned to the DC element, and a relatively lower weight is assigned to the first few chosen AC elements. The AC elements are arranged in a zigzag manner by the DCT, and the AC elements closer to the DC element have a higher importance than the AC elements further apart. A modification in an image block will be highly noticeable through the DC element but also through the AC elements closer to the DC element. The elements further apart may not capture the changed image block as much.

By assigning a higher weight to the DC elements, they are protected with a larger sub-MAC as compared to the sub-MAC length chosen for the AC elements. As the DC element has a higher weight, the number of iterations of SID used for correcting the modifications in the DC element is higher than the number of iterations used for correcting the corresponding group of AC elements. It is however more important that the DC element is correct or corrected as compared to the corresponding group of AC elements. For simplicity, it is assumed that the length of a WNTMAC tag is equal to the length of an NTMAC tag. This is ensured by compensating the longer length of the sub-MAC for the DC elements through a smaller length sub-MAC for the AC elements.

7.6 Simulation Setup

Simulation results are given in this chapter in order to show the effectiveness and a comparison of the algorithms discussed above. Simulation results for both of the algorithms discussed above are based on the transmission of images over an AWGN channel with BPSK modulation. Turbo codes of rate-1/3 are chosen as the channel codes. Simulation results are also given in the absence of any channel codes in which case the channel measurements are used to obtain reliabilities. The LLR values produced by the decoder in case of Turbo codes and the channel measurements in case of no codes are used for estimating the bit reliabilities.

The source image is chosen to have a resolution of 128×128 pixels, and each pixel value is represented by 1 byte. If a coloured image is considered, then it must be converted internally to greyscale before the NTMAC/WNTMAC tags can be computed on it. A block size of 8×8 pixel is chosen, and therefore for the given image resolution, there are 16 blocks in each row and 16 blocks in each column. The blocks are nonoverlapping; therefore a total of 16×16 blocks are considered in the simulations. SHA-256 is used for calculating the MAC tag. It is assumed that the keys are pre-shared, i.e. key exchange is not considered in this text. A sub-MAC length of 128 bit is chosen. Thus, for 16 blocks in a row, the NTMAC/NTMAC length will be $128 \times 16 = 2048$.

After calculating DCT on each image block, the DC elements are protected using one of the algorithms as explained earlier. The DC/AC elements require 16 bits for storage, i.e. 2 bytes. For the case of WNTMAC, there will be two tags for each row or column of blocks, one corresponding to the DC elements of the row of block or column. The next WNTMAC tag is computed on the corresponding AC elements, and it is shorter in length as compared to the WNTMAC tag for the DC elements.

The number of SID iterations is chosen to be 216 for the WNTMAC corresponding to the DC elements, whereas it is chosen to be 28 for the WNTMAC corresponding to the AC elements. With a higher number of iterations and therefore a larger number of least reliable bits considered in SID, there is a higher probability of data correction. Thus a higher number of bits of the DC elements can be corrected effectively correcting the corresponding image block. On the other hand, the lower number of iterations used for the AC elements will consider only half of the least reliable bits. Thus the probability of correcting the corresponding image block through the AC elements is lower. However, it is desirable to correct the DC elements more than the corresponding AC elements. From the authentication perspective as well, the DC elements are more important than the AC elements, and therefore the concept of relative weights is chosen here.

7.7 Simulation Results

Figure 7.1 shows the simulation results for ELC-NTMAC when the Lena image is transmitted over AWGN channel with BPSK modulation. The results are shown in the absence of any channel codes at an SNR of 11 dB. Figure 7.1a shows the original Lena image. Figure 7.1b shows the original Lena image in which the modified blocks are marked using a black block. Figure 7.1c shows the identified suspicious blocks using the NTMAC tags. The suspicious blocks are marked as red. It can be seen that all the modified blocks are identified by the NTMAC tag as well as a few

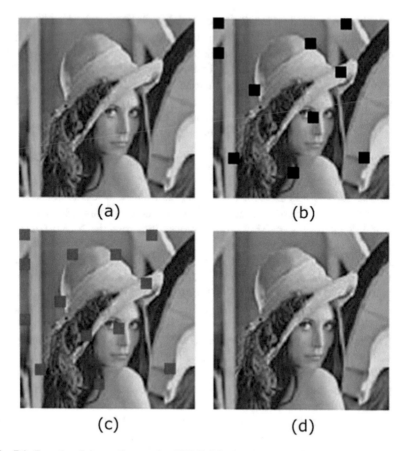

Fig. 7.1 Error localizing and correcting NTMAC in the absence of channel codes. (**a**) Original image. (**b**) Image with actual modified blocks. (**c**) Image with suspicious blocks marked as red. (**d**) Modifications corrected with ELC-NTMAC

additional blocks. The additional blocks are identified as suspicious as the modifications might have occurred in the sub-MAC of the NTMAC tag rather than the image blocks themselves, which the ELC-NTMAC algorithm perceives as modifications in the image block because of the mismatch of the sub-MAC. Figure 7.1d shows that the identified suspicious blocks have been corrected using the iterative decoding of the SID algorithm.

Figure 7.2 shows the simulation results for Lena image, when Turbo codes of rate-1/3 are used as compared to Fig. 7.1 where no codes were used. The results are shown for an SNR value of 2.5 dB. The explanation is similar to that for Fig. 7.1.

Figure 7.3 shows the simulation results for ELC-WNTMAC for Lena image, when no channel codes are used at an SNR value of 11 dB. The suspicious blocks identified in Fig. 7.3c are corrected by the algorithm as shown in Fig. 7.3d.

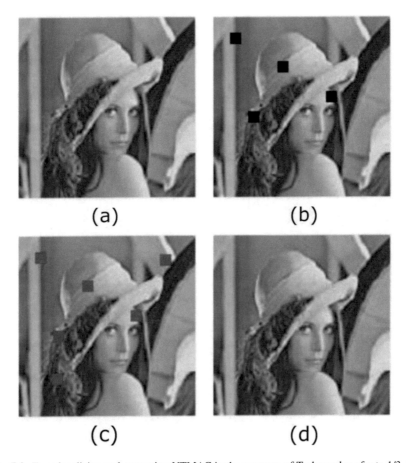

Fig. 7.2 Error localizing and correcting NTMAC in the presence of Turbo codes of rate-1/3. (**a**) Original image. (**b**) Image with actual modified blocks. (**c**) Image with suspicious blocks marked as red. (**d**) Modifications corrected with ELC-NTMAC

Figure 7.4 shows the simulation results for ELC-WNTMAC for Lena image, when Turbo codes of rate-1/3 are used as compared to Fig. 7.3 where no channel codes were used. The results are shown for an SNR value of 2.5 dB. The suspicious blocks identified with the help of WNTMAC are shown in Fig 7.4c which are corrected using the ELC-WNTMAC algorithm as shown in Fig 7.4d.

Figure 7.5 shows the simulation results for ELC-WNTMAC for Lena image in the presence of forgery attack. Figure 7.5a shows the original Lena image. Figure 7.5b shows a forged image, where extra hair are visible on the forehead. Figure 7.5c shows that the modified area is identified using the ELC-WNTMAC algorithm. Figure 7.5d shows that the forged area is not correctable; however, the only block marked as suspicious aside from the forged block has been corrected. The forged cannot be corrected due to the fact that it is quite larger than the error correction capability of the ELC-WNTMAC. However, the forged area is identifiable using the algorithm.

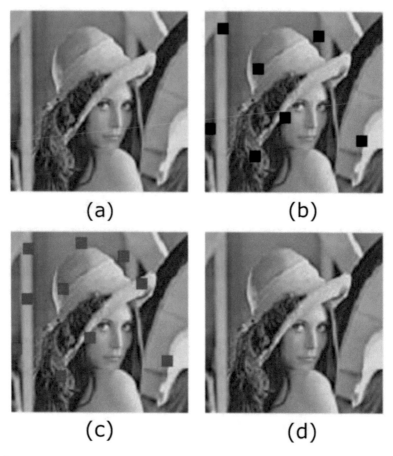

Fig. 7.3 Error localizing and correcting WNTMAC in the absence of channel codes. (**a**) Original image. (**b**) Image with actual modified blocks. (**c**) Image with suspicious blocks marked as red. (**d**) Modifications corrected with ELC-WNTMAC

7.8 Conclusion

In this chapter, two algorithms for image authentication are discussed. The algorithms are not only able to authenticate images in the presence of noise but also able to localize errors. The localization is at a block level. The granularity of block size is controllable and can be adjusted as desired. The algorithms discussed in the chapter are also able to correct certain modifications in addition to locating them. Simulation results at various noise levels, in the presence of unintentional modifications as well as forgery attacks, are given.

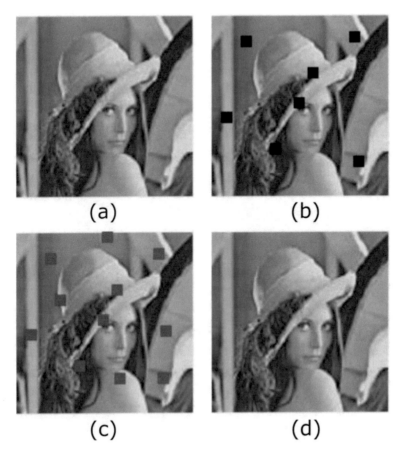

Fig. 7.4 Error localizing and correcting WNTMAC in the presence of Turbo codes of rate-1/3. (**a**) Original image. (**b**) Image with actual modified blocks. (**c**) Image with suspicious blocks marked as red. (**d**) Modifications corrected with ELC-WNTMAC

Fig. 7.5 Error localizing and correcting WNTMAC in the presence of forged image. (**a**) Original Lena image. (**b**) Forged image. (**c**) Detection of forged area marked through red blocks. (**d**) Forgery is not correctable using ELC-WNTMAC but erroneous block is corrected

References

1. P.G. Freitas, R. Rigoni, M.C. Farias, Secure self-recovery watermarking scheme for error concealment and tampering detection. J. Braz. Comput. Soc. **22**(5), 5 (2016)
2. Y. Wu, Tamper-Localization Watermarking with Systematic Error Correcting Code. in *International Conference on Image Processing,* Atlanta, GA, 8–11 October 2006, pp. 1965–1968
3. M. Chen, Y. He, R.L. Lagendijk, A fragile watermark error detection scheme for wireless video communications. IEEE Trans. Multimedia **7**(2), 201–211 (2005)
4. S.A.H.A.E. Tabatabaei, O. Ur-Rehman, N. Zivic, AACI: The mechanism for approximate authentication and correction of images. in *IEEE International Conference on Communications Workshops (ICC),* Budapest, 9–13 June 2013, pp. 717–722
5. N. Zivic (ed.), *Robust Image Authentication in the Presence of Noise* (Springer, New York, 2015), pp. 43–74, ISBN 978-3-319-13156-6, https://doi.org/10.1007/978-3-319-13156-6_2

Index

A

Additive white Gaussian noise
(AWGN), 18, 58, 59
Approximate message authentication code
(AMAC), 23–25, 45, 49
Authentication
code frameworks, 62
and correction of images, 62
MAC tag, 54
RoI, 56, 58
SID, 53
watermarking method, 61
Automatic repeat request (ARQ), 15

B

Base stations (BSs), 8
Bit error rate (BER), 19, 58, 59
Block level localization, 63, 70

C

Classification, digital watermarking
compression, 37
embedding, 35–36
perceptibility, 37
robustness, 36
Content authentication
vs. data, 5
standard mechanisms, 6

D

Data authentication
encryption schemes, 2
properties, 2
standard method, 2
Decoding
and authentication, 55
RoI, 57
two-step authentication, 58
Diffraction, 8
Digital Video Broadcast (DVB-S2), 18
Digital watermarking
capacity, 35
carrier signal, 33
compression, 37
description, 33
embedding, 34–36
extraction, 34
generation, 33–34
invisibility, 34
perceptibility, 37
requirements, 37
robustness, 35, 36
tamper detection, 35
Discrete cosine transform
(DCT), 36, 40–42, 62
Discrete wavelet transform
(DWT), 36, 47, 48
Doppler shift, 15, 16
Dual watermarking, 40, 46, 51

© Springer International Publishing AG, part of Springer Nature 2018
O. Ur-Rehman, N. Zivic, *Noise Tolerant Data Authentication for Wireless
Communication*, Signals and Communication Technology,
https://doi.org/10.1007/978-3-319-78942-2

Printed in the United States
By Bookmasters